The
NATURE
of REALITY

The
NATURE
of REALITY

Richard Morris

McGraw-Hill Book Company

New York   St. Louis   San Francisco   Hamburg   Mexico   Toronto

1 2 3 4 5 6 7 8 9 DOCDOC 8 7 6

ISBN 0-07-043278-3

LIBRARY OF CONGRESS CATALOGING-IN-PUBLICATION DATA

Morris, Richard.
  The nature of reality.
  1. Reality.   2. Matter.   3. Cosmology. I. Title.
BD331.M764 1986   110   86-7106
ISBN 0-07-043278-3

BOOK DESIGN BY PATRICE FODERO

Contents

Bizarre Theories and the New Physics

Anyone who has read accounts of the "new" physics that has grown up since about the mid-1970s can easily gain the impression that physicists have a tendency to engage in a lot of bizarre speculation nowadays. One reads, for example, of theories based on the postulate that there are not three dimensions of space, but rather nine or ten. There are theories describing the events that supposedly took place when the universe was less than a billionth of a billionth of a billionth of a second old and had a temperature of some ten billion billion billion degrees. One sometimes hears talk of antigravitational forces, or of shadow matter, a hypothetical substance which no one will ever be able to see but which might be the primary constituent of the universe. There are theories which suggest that the universe may have popped into existence out of nothing, and that all the matter and energy that it contains may have been created in the ensuing expansion. There is yet another theory which is based on the idea that the universe splits into innumerable copies of itself at every moment, and that an infinity of "parallel" universes may now exist.

It is not always easy to tell just what one should make of all this. Indeed, the lay person who wonders whether physicists have ceased to study observable phenomena to engage in a program of speculation about fantastic ideas cannot be belittled for his doubts. Physicists sometimes do appear to have left reality behind and to have gone off in search of bizarre new theoretical concepts.

Sometimes one's first reaction to new theories is to say to the theoretical physicists, "You must be crazy." But of course they are not. The reason that many of the theories that constitute the so-called new physics have "bizarre" elements is that conceptions of reality are changing. Old ideas are being replaced by new ones that often appear to be "crazy" simply because they are unfamiliar.

As science enters new territory, the structure of scientific thought must change. Old conceptions must give way to new ones, and new theoretical outlooks must be developed. During periods of intense experimental and theoretical activity, scientific ideas about the nature of physical reality are transformed. There is no way that this can be avoided. A coherent picture of nature cannot be created from outmoded concepts.

If some of the new theories seem bizarre, perhaps we should remember that many of the ideas that we all accept today seemed crazy at one time. The notion of chemical elements must certainly have seemed crazy to those who had grown up with the old notion that all material substances were combinations of earth, air, fire, and water. The idea that stones could fall from the sky was once considered to be a crackpot belief. Today, we all know that meteorites commonly plunge to the surface of the earth. At one time, astronomers refused to consider the idea that there were galaxies other than our own. Our Milky Way galaxy, they said, *was* the universe. The nebulae that were observed in the night sky were said to be nothing but glowing clouds of interstellar gas. When it was first suggested that these nebulae were other "island universes," the idea was thought to be bizarre.

Many twentieth-century ideas have also appeared to be too crazy to be true when they were first proposed. When some of Albert Einstein's theories were first published, they encountered ridicule in some quarters. In some cases, they were accepted only grudgingly. When the members of the Swedish Academy voted to give Einstein the Nobel Prize in physics for 1921, they were apparently afraid to give it for such a bizarre idea as relativity. Relativity was never mentioned in the citation that accompanied the prize; Einstein was given the award for his contributions to quantum theory instead. On another occasion, an eminent physicist recommended Einstein for a scientific position. He seems to have felt that he had to apologize for some of the latter's "crazy" ideas; he said of one of Einstein's theories (the one for which

Einstein later received the Nobel Prize) that it could not "really be held against him."

It is possible to cite numerous similar examples. Today, virtually every literate person has heard of black holes. The concept is so familiar that it is even alluded to in television commercials. But when the idea of a black hole was first proposed, physicists thought it so bizarre that they would not even consider it. They would not believe that a dying star could become so compressed that nothing, not even light, could escape from its intense gravitational field. An idea that is at least as familiar as that of a black hole is that of the big bang that took place shortly after the creation of the universe. But at one time, physicists refused to take this idea seriously either; they did not believe that it was possible to say anything meaningful about an event that had presumably taken place billions of years in the past. In this case, they didn't even bother to respond with ridicule. The theory was simply ignored.

If physicists have had a tendency to resist revolutionary new ideas, perhaps it is not surprising that lay people are often baffled by them, particularly when they have expended some effort learning to understand the old ideas that the new theories are intended to replace. It is not easy to change the way one thinks about the universe. The fact that old scientific ideas have a way of becoming part of "common sense" once they have been widely disseminated makes the process of change even more difficult.

Yet just such a change is necessary if one wants to understand what the new physics is all about. Since the 1960s and 70s the pace of change has accelerated. Furthermore, the changes that have taken place have not been limited to the promulgation of new theoretical ideas. The fundamental concepts upon which theoretical structures are built have also been modified. Today, scientists' conceptions of such basic ideas as "space," "time," "force," and "matter" are changing. Changes in these basic ideas have led, in turn, to modifications in the way physicists think of such familiar entities as the electron. "Electron" is a term that scientists have been using since the 1890s. However, when the term is used today, it does not have quite the same meaning that it had in 1910, or even in 1950.

In modern theories, such elementary particles as the electron, the proton, and the neutron are not viewed the way they were a few years ago. In fact, many physicists now believe that particles are not fundamental constituents of matter. According to these scientists, the physical world is made up of quantum fields, which can manifest themselves as particles in many different ways. Quantum fields are all that exists, they say; there is nothing else.

In some of these field theories, the traditional distinctions between such apparently dissimilar entities as force and matter are beginning to disappear. According to quantum field theory, force and matter can be associated with two different kinds of particles produced by quantum fields. However, in some of the more advanced field theories, the distinction between these particles has become blurred.

The development of these and other ideas has led to changes in the ways scientists view the universe as a whole. For example, new (and as yet unconfirmed) theories suggest that it might be possible for a strange substance called shadow matter to exist. Shadow matter is matter that would interact with ordinary matter only gravitationally. As a result, if it exists, it could neither be felt nor seen. Light and the solidity of matter to our touch both depend upon nongravitational forces. Consequently, one could stand on a shadow matter mountain and never realize it was there.

According to one hypothesis, shadow matter might constitute the bulk of the matter in the universe. Since about 1975, it has been discovered that approximately 90 percent of the mass in the universe exists in the form of a mysterious dark matter which fills interstellar and intergalactic space. It is not known what the constituents of this dark matter are. However, the idea that it is shadow matter is one of the more intriguing (and speculative) hypotheses. There are, of course, less "bizarre" possibilities. Many physicists believe that the dark matter might be made up of exotic particles that were created in the hot fireball that resulted from the big bang.

Speculation about the events that took place shortly after the creation of the universe has led to new theories about the universe and about the nature of matter as well. Physicists now realize that the universe may originally have contained no energy or matter, that both may have been created during a stage of rapid expansion of the universe shortly after it was formed. They also hope that theories which deal with the early universe will help them to better understand the nature of quantum fields. The particles that were created in the big bang fireball possessed energies much greater than those which can be created in the most powerful particle accelerators. Thus, strange as it may seem, the big bang has become a kind of laboratory for high-energy physics.

As physicists have developed theories that peer farther and farther into the past, they have begun to speculate about the events that may have caused the universe to come into existence. At one time, this was thought to be an area that was "beyond physics." It is not any longer. New kinds of speculation about the nature of space and time have led to hypotheses about their creation. Three-dimensional space and four-dimensional space-time are no longer taken for granted. It is understood that there might have been a "time" when neither space nor time existed. Alternatively, there may have been time before there was any such thing as space.

Physicists now realize, also, that space may have more than three dimensions. When the universe was created, there may have been nine or ten.* If there were, then there are still nine or ten today. The reason that we are aware of only three is that the extra ones have presumably collapsed to dimensions (here I am using "dimensions" in the sense of "size") far smaller than those of an atomic nucleus. This raises the possibility that other kinds of collapse might be possible and that a universe with a dimensionality different from that of ours is not beyond the realm of possibility.

If such universes are possible, they might very well exist. At one time, it was believed that there was only one universe—the one in which we live. To hypothesize about the possible existence of other universes was thought to be a kind of speculation that was properly left to science fiction writers. This is true no longer. Theories which invoke the existence of other universes have become commonplace, and some physicists have speculated that the number of universes might be infinite.

* Which makes ten or eleven dimensions in all if the time dimension is taken into account. Thus physicists habitually speak of ten or eleven dimensions of space-time.

Some of the discoveries that have been made in recent years are so strange and so surprising that physicists who have tried to work out the implications of these discoveries have found themselves speculating about matters that were once thought to be the province of metaphysically inclined philosophers. They have found themselves asking questions like the following:

Why is there something rather than nothing?

Does physical reality contain everything that *can* exist, or is it made up only of those things that *must* exist?

Are there really laws of nature, or are they only constructs of the human mind?

Is a universe that does not harbor conscious beings possible?

To what extent can we really say that there is such a thing as objective reality?

Some of these questions have arisen in the course of debate about the meaning of quantum mechanics, the theory upon which most of the new physics is based. Although quantum mechanics was developed in the mid-1920s, physicists have not yet reached agreement about all of its implications.

In more recent years, the debate has intensified. New theoretical discoveries, such as the "many-worlds" interpretation of quantum mechanics and a mathematical result known as Bell's theorem, have led to heightened speculation about the nature of quantum reality. Some physicists have suggested that quantum particles might be able to exert influences on one another that travel at velocities greater than that of light. Others speak of the existence of numerous parallel quantum universes. Yet others conclude that quantum mechanics implies that reality depends upon the existence of sentient observers, that if we or beings like us did not exist, then the universe would have no reality either.

New ideas and new conceptions have been produced at such a prodigious rate that it is sometimes difficult to distinguish between scientific fact, scientific speculation, and philosophical inquiry. As a result, the contemporary reader of books on the new physics is often either left bewildered or given the impression that certain speculative ideas have become more generally accepted than they really have.

I don't think that the authors of such books really intend to mix fact and speculation in this way. I suspect that the reason they often do so inadvertently is that like the rest of us, scientists have a tendency to get caught up in their ideas. As a result, unproven hypotheses sometimes become so appealing to them that they find it difficult to imagine that these hypotheses could eventually be proved wrong. Thus they have a tendency to commit themselves a little too strongly to ideas that have not yet been tested by experiment.

Speculation plays an important role in science. Before one can discover what is true, it is necessary to find out what might be true. Hence it would be impossible to write a book about the new physics by discussing only well-established results. If an author were to ignore speculative ideas, his account would be just as misleading as it would be if he discussed only that which was speculation.

In the chapters that follow, I will attempt to distinguish between generally accepted results and ideas of a speculative nature. I will not refrain from discussing the latter, however. Nowadays, scientific ideas change rapidly. Today's wild speculation is often tomorrow's experimentally confirmed theory. Any book that did not consider the speculative would be out of date by the time that it was published.

I think that the reader will find that, even if we consider only reasonably well established fact, the results achieved by the scientists who are creating the new physics are astonishing indeed. Scientific paradigms are changing to such an extent that new kinds of understanding of the nature of reality are beginning to emerge. There is much that physicists do not yet understand about the physical world, but it is beginning to be clear that reality is not quite what they thought it to be even in the 1960s and 70s.

Part One

The Nature of Matter

1

The Nature of the Electron

Since physicists are always attempting to push back the frontiers of knowledge, perhaps there is nothing very surprising about the fact that they often find themselves confronted by scientific paradoxes. Since the early twentieth century, there has rarely been a time when they have not found themselves wrestling with unanswered questions as they sought to understand the implications of the theories that they had developed. Some of these theories seemed so strange at first that they were rejected by the majority of working scientists. And yet these strange theories have often yielded results which have been confirmed by experiment to an amazing degree of accuracy. That arbiter of truth that we call "experiment" has forced the acceptance of new outlooks and new interpretations of the nature of physical reality.

Even today there are theories which contain features that no one understands but which are so well confirmed that no one would dream of giving them up. That such a situation should exist should not be considered odd. Physics has had a tendency to get ahead of itself since the beginning of the twentieth century. There have frequently been theories which proved to be successful even though no one really knew why they worked. Physicists have made experimental discoveries that were explained only decades later. In spite of its great successes, modern physics has always seemed to contain a residue of theoretical and experimental results that just don't seem to make any sense.

Perhaps this is a measure of physicists' success rather than an indication of failure. After all, if everything were understood, science would immediately come to an end. If there were no puzzles, it would be impossible to carry out research. The very fact that unanswered questions exist gives scientists something to investigate.

Thus one should not be surprised to discover that the electron—one of the constituents of atoms—was discovered at a time when a number of prominent physicists doubted that atoms existed, and when others commonly defined the atom as the smallest particle of matter. It seems somehow appropriate that the electron should have seemed somewhat paradoxical from the beginning. Many of the strange results that modern physicists have obtained have arisen out of attempts to understand this puzzling particle.

When the British physicist J. J. Thomson* announced his discovery of the electron in 1897, many scientists refused to take him seriously; some suspected that he was playing a kind of practical joke. But the ridicule subsided when further experimentation demonstrated that the existence of electrons was a fact that had to be accepted, no matter how odd it seemed. Thomson's "ridiculous" results rapidly came to appear significant indeed. Thomson received the Nobel Prize for his work in 1906, and in 1908 he was knighted.

Today Thomson is remembered as one of the great scientists. Yet in Thomson's own time no physicist in his right mind would take Thomson's conception of the electron seriously. Since then, ideas about this tiny particle have changed so dramatically that it

* His full name was Joseph John Thomson, but his contemporaries knew him as "J.J." Physicists today continue to refer to him as simply "J. J. Thomson."

is possible to say that there is little similarity between Thomson's electron and the electron as it is viewed by modern physicists.

As new discoveries are made, it is not just theories that change. Ideas about fundamental concepts and about the basic constituents of matter change also. If J. J. Thomson somehow managed to return from the dead and listened to contemporary physicists discussing the nature of the electron, it is safe to say that he would not know what they were talking about.

But this does not imply that his discoveries were not significant. That would be like saying that Magellan's circumnavigation of the globe between 1519 and 1522 was a meaningless feat because one can travel around the world so much faster today. Even though Thomson's conception of the electron was eventually discarded, his discovery was extremely important. Thomson's work led to the discovery of a seemingly endless series of new puzzles. He found few enduring answers, but his work made physicists realize that there were a multitude of unanswered questions. And finding possible solutions to unanswered questions is what physics is all about.

In order to understand the significance of Thomson's work, it might be best to backtrack a bit and to make a few remarks about the state of atomic theory when he made his discovery. In order to do this, it will be necessary to make a few brief comments about Thomson's predecessors. It would be difficult to understand just what it was that he did if one knew nothing about the outlooks that were common in his day.

Although the existence of atoms was first proposed by the Greek philosophers Leukippos and Demokritos during the fifth century B.C., modern atomic theory can be said to have originated with the work of the English chemist John Dalton. In 1803, Dalton pointed out that many of the facts of chemistry could be explained if one assumed that all of the chemical elements were made up of tiny, indivisible particles called atoms. According to Dalton, these particles were the smallest constituents of matter; all substances were made up of various kinds of combinations of these atoms.

Dalton's theory was rapidly accepted by chemists. During the nineteenth century, numerous physicists, too, became converts when they discovered that atomic theory could be used to explain such phenomena as the conduction of heat and the behavior of gases. Yet, at the beginning of the twentieth century, there were still a number of influential scientists who continued to express doubts about the atomic hypothesis. One of these was the Austrian physicist and philosopher Ernst Mach, whose writings were later to be influential upon the work of Albert Einstein. Another skeptic was the Russo-German physical chemist Wilhelm Ostwald, who would receive the Nobel Prize in chemistry in 1909.

According to Mach and Ostwald, atoms were nothing more than a useful fiction. It was true that the hypothesis of their existence could be used to explain various different kinds of physical and chemical phenomena. But this did not necessarily imply that they were real. As Ostwald put it, "At best there follows from this the *possibility* [his italics] that they are in reality; not, however, the *certainty*." Mach's view was even more extreme. Not only did he believe that atoms were nothing but hypothetical contrivances, he also formulated a philosophy of science that put all theoretical statements into this category. According to Mach, all of the concepts of physics had only a hypothetical kind of reality. In his view, they were much less "real" than direct perceptual data such as sounds and sensations of color.

The Mach-Ostwald view was typical of a philosophical outlook that is generally called *positivist*. Adherents to the doctrine of positivism regard all abstract ideas as constructs; they say that we should only attribute reality to that which we can perceive.

It is not my intention to enter upon a critique of the positivist movement in philosophy. I will confine myself to noting that Mach's and Ostwald's arguments did not seem to be so very unreasonable to nineteenth-century scientists. After all, atoms could not be seen, not even through the most powerful microscopes. Thus the evidence for their existence was of an indirect nature. The primary reason for believing in the existence of atoms was that the theories that were based on the atomic hypothesis seemed to work—under most circumstances, at any rate. But, like most theories, they also had their failures. For example, although the hypothesis of the existence of atoms explained many of the observed facts about the behavior of heat, the theory failed when physicists tried to use it to calculate how much heat a body would absorb.

Thus when Thomson carried out his experiments on electrons, he was working in a climate in which an attitude of skepticism still seemed perfectly reasonable. It is easy to see why some of his contemporaries thought his results bizarre. Even though the existence of atoms had not yet been experimentally demonstrated, Thomson was claiming to have discovered that there were particles that were even smaller.

When Thomson embarked on his research, his intent was not to show that atoms had constituents, or even that they really existed. He was simply studying electrical discharges which could be made to take place in glass tubes from which most of the air (or other gaseous contents) had been evacuated. The study of electric currents was one of the major preoccupations of nineteenth-century physics. Thomson initially intended to do no more than investigate some phenomena associated with the passage of electricity through gases. At the time, some of these phenomena were considered to be very puzzling.

In 1859, the German mathematician and physicist Julius Plücker had discovered that when a vacuum pump was used to remove most of the gas that had been used to fill a tube, an electric current could easily pass through the gas that remained. In 1879, the British physicist Sir William Crookes observed that when the pressure was very low and the gas very tenuous, a curious phenomenon could be observed. A fluorescent glow would appear on the glass at one end of the tube. Something was apparently being emitted from the *cathode,* or negative electrode.

Although numerous experiments were performed by Crookes and by other physicists in the years that followed, no one was able to determine just what these *cathode rays* were. In fact, a controversy concerning their nature soon arose. The British physicists who studied them tended to think that they were particles of some kind. The German physicists, on the other hand, thought them to be a type of radiation. The German physicist Heinrich Hertz, for example, discovered that the "rays" could be made to pass through thin films of metal. When the films were subsequently examined, no puncture marks could be seen. Cathode rays, Hertz concluded, were obviously some new form of light.

British physicists performed experiments that seemed to show that cathode rays had a negative electrical charge. In their view, this implied that the "rays" must be particles. Hertz countered with an experiment of his own. He passed cathode rays between a pair of electrically charged plates, and found that no deflection could be observed. If cathode rays really had a negative charge, he pointed out, they would have been attracted to the positively charged plate.

Like his British colleagues, Thomson suspected that cathode rays were streams of particles. But if he wanted to show that this hypothesis was correct, he had to discover what was wrong with Hertz' experiment. Suspecting that the electrical discharge in the tube had somehow neutralized the charges on Hertz' plates, Thomson repeated the experiment, using a tube that had been pumped down to a near vacuum. He found that when enough gas was evacuated from his apparatus, the electrical deflection could be observed.

The only reasonable interpretation of this result was that cathode rays were indeed made up of electrically charged particles. Furthermore, the charge of these particles was negative, since they were deflected toward positively charged plates. Cathode ray particles undoubtably existed, but Thomson did not know very much about them. He did not know how big they were, how much they weighed, or even the amount of charge that each particle carried. Nor had Thomson been able to determine the velocity of the particles to any great degree of accuracy. He only knew that their motion was very rapid.

In order to find out more about the nature of the cathode ray particles, Thomson devised an ingenious experiment. He constructed* a tube containing a coil which produced a magnetic field and also containing the usual electrically charged plates.

Charged particles are deflected both by magnetic fields and by

* Actually, it was Thomson's assistants who made the apparatus and who performed most of the experiments. Although Thomson had a remarkable instinct for knowing what problems were worth working on, and although he exhibited considerable ingenuity in devising experimental methods, he was rather clumsy when it came to manipulating laboratory instruments. Indeed, it was often said that it was best not to let Thomson get too near one of his experiments. However, working under Thomson must not have been a bad kind of training. Some seven of his research assistants went on to receive Nobel Prizes for discoveries of their own.

the electric fields which exist between a pair of plates that are positively and negatively charged. Thomson realized that if he was able to balance the two fields so that their effects canceled one another out, he would be able to determine the particles' velocity.

The reason he could do this is that the force which a magnetic field exerts on a particle depends on both the particle's electric charge and its velocity. On the other hand, the force exerted by an electric field depends on charge alone. If one adjusts an apparatus like Thomson's so that the two fields have equal and opposite effects, then the quantity of charge becomes irrelevant, and the particles' velocity can be calculated from measurements of the electric and magnetic field intensities.

Once Thomson had calculated the velocity of the particles, it would have been a simple matter to determine their charge, if he had only known how much the particles weighed. All he would have had to do would be to turn off the magnetic field and measure the deflection caused by the electrically charged plates. Knowing the deflection, he could have computed the force exerted on the particles and consequently their charge.

However, Thomson could devise no way of determining the particles' mass. Consequently, the best he could do was to compute a ratio of mass to charge. He discovered that this ratio was more than a thousand times less than that which was characteristic of *ions*, or charged atoms. It appeared that there were two possibilities: either a cathode ray particle weighed much less than an atom, or it weighed about the same and had a much larger charge.

Of the two possibilities, Thomson considered the former to be more reasonable. He theorized that the particles he had studied were considerably less massive than atoms. Of course, he was correct. Modern experiments have established that an electron is 1836 times lighter than a hydrogen atom, for example.

Thomson was not the first to use the term "electron." This was a name that had previously been proposed by the Irish physicist George Johnstone Stoney for the then hypothetical "atoms of electricity." Further experiments convinced Thomson that the electron must be a universal constituent of matter. When he experimented with tubes that contained different kinds of gases, his results were always the same. Thomson concluded that if all gas atoms contained the same kind of electron, then it was reasonable to assume that electrons were constituents of the atoms that made up all chemical elements.

Ordinarily, atoms are electrically neutral. Thomson concluded that if atoms contained negatively charged electrons, then they must also contain an equal amount of positive charge. In 1904, he proposed what later came to be known as the "plum pudding" model of the atom. According to Thomson's theory, atoms were made up of spheres of positively charged material in which the electrons were embedded. The electrons were the "plums" and the positive matter the "pudding." According to the theory, electrons were quite complicated objects. Thomson thought (incorrectly) that each atom contained thousands of electrons.

Thomson's theory about the structure of the atom must have seemed quite audacious to many of his contemporaries. After all, it had not yet been conclusively demonstrated that atoms even existed. However, this little difficulty was soon eliminated, by the French physicist Jean Perrin, and by an unknown German physicist named Albert Einstein.

At about the same time that Thomson was working out his theory of atomic structure, Einstein began to wonder if there might not be a way to demonstrate that molecules—and consequently atoms—really existed. He noted that no one had yet come up with an adequate theoretical explanation of a phenomenon called *Brownian movement,* and soon realized it might provide a clue to the nature of reality on the microscopic level.

In 1827, the Scottish botanist Robert Brown had happened to examine a drop of water which contained some minute pollen grains. When Brown looked at the pollen through a microscope, he noticed that the grains moved about continuously in a random way. Something was causing them to zigzag about, first in one direction, and then in another. One thing that he found especially puzzling was the fact that this movement never stopped.

At first, Brown's discovery did not attract much attention. When physicists finally did take notice of it, they attempted to explain the phenomenon as something that was caused by the absorption of light or the conduction of heat through the liquid. If these explanations were not very convincing, no one really noticed. After all, Brownian movement did not seem to be a very significant phenomenon.

Brownian movement was left unexplained until 1905. In that year—the same year in which he published his special theory of relativity—Einstein published a paper in which he showed that Brownian movement could be accounted for if one assumed that the motion of the particles was caused by collisions with the molecules of the liquid. According to Einstein, since molecules presumably moved about in a random manner, the collisions would cause particles that were suspended in a liquid to move in an irregular pattern also. However, Einstein pointed out, the motion was not quite as irregular as it seemed. It was possible to calculate what the average deflection of one of the particles should be. Furthermore, it should be possible to make use of measurements of these deflections to determine the size of molecules, and hence of atoms.

Einstein's analysis was ingenious, but it seemed at first that physicists would be forced to conclude that it was incorrect. Three different experimental studies, one of which was carried out before Einstein's paper was even published, contradicted his results.

Then, in 1908, Perrin performed a series of more meticulous experiments. Not only did he confirm that Einstein's theory was correct, but he was also able to determine the number of atoms in a given quantity of matter. This turned out to be about 10^{22} atoms per gram.*†

Perrin's results could be used to determine the mass of an atom, but not of an electron. Thus Thomson's mass-charge ratio was still the only quantity that was known to any degree of accuracy. However, this problem was soon cleared up also. During the years 1908 to 1913, the American physicist Robert Millikan carried out a series of experiments in which he made accurate determinations of the electron's charge. This made it possible to compute the electronic mass also. The mass of the electron turned out to be 9.8×10^{-28} grams.‡

* 10^{22} is the number represented by the numeral 1 followed by twenty-two zeros.
† The number of atoms per gram varies, depending upon the substance being studied. Perrin actually expressed his results as atoms per *gram-mole*. "Gram-mole" is a term used by chemists that is related to molecular weight.
‡ 10^{-28} is the number 1 divided by 10^{28}.

By around 1910, it appeared that all of the problems concerning the nature of the electron had been solved. One could think of it as a tiny sphere that was a constituent of the atom. Accurate determinations of its charge and mass were beginning to be made. Nor did the behavior of electrons seem to be very mysterious. In fact, electrons could be used to explain the emission of light from atoms. If Thomson's theory was correct, then it seemed to follow that electrons would vibrate back and forth within an atom. If the velocity of the vibrating electrons changed, some energy would have to be given up or absorbed. Since light was a form of energy, it seemed that Thomson's theory explained both the emission and absorption of light by material substances.

In 1911, the British physicist Ernest Rutherford, one of Thomson's former assistants, showed that there was something terribly wrong with this picture. Bombarding atoms with alpha particles* that were emitted from the newly discovered radioactive elements, he found that Thomson's plum pudding model could not possibly be correct.

Rutherford directed beams of alpha particles at sheets of gold foil. After the particles passed through the foil, they were allowed to strike a fluorescent screen. The resulting impacts produced small flashes of light that could be seen through a microscope. Observation of these flashes allowed Rutherford and his assistants to measure the angles through which the paths of the alpha particles had been deviated when they passed through atoms of gold in the foil.

Rutherford found that, as he had expected, the deflection of the majority of the alpha particles was slight. But a small number of the particles were deflected through large angles, and about one in twenty thousand turned right around and rebounded back in the direction of the alpha particle source.

Rutherford was later to describe his discovery—that such rebounds took place—as "the most incredible event that has happened to me in my life." It is not hard to see why he was so surprised. According to then accepted ideas about the atom, such a phenomenon was impossible. It was obvious that the large de-

---

* An alpha particle is made up of two protons and two neutrons. It is identical to the nucleus of a helium atom.

flections could not have been caused by collisions between the heavy alpha particles and the very light electrons; an alpha particle would simply push any electrons it encountered out of its way. Nor could the alpha particles be deflected in this manner by Thomson's sphere of positive charge; they would plow right through it, and their trajectories would change by only small amounts. The observed rebounds of a few alpha particles, Rutherford realized, could be explained in only one way. The positive charge in an atom was not spread out over a relatively large sphere; it had to be concentrated within a small nucleus in the atom's center. The nucleus had to be small because only a few of the alpha particles seemed to encounter it.

Rutherford's discovery of the atomic nucleus raised more questions than it answered. If negatively charged electrons surrounded a positively charged nucleus, then it seemed reasonable to suppose that an atom must be a kind of miniature solar system. The electrical attraction between the positive and negative charges would cause the electrons to orbit the nucleus in a manner analogous to that in which gravity causes planets to revolve around the sun.

This picture of the atom as a miniature solar system seems to have been unusually persistent. As we shall see, it is not a particularly accurate model. Even in 1911, it was obvious that there was something wrong with it. It was immediately apparent, to Rutherford and to the other physicists of the day, that if the planetary model was correct, then atoms could not exist. A simple calculation showed that orbiting electrons would spiral into the nucleus in less than a hundred-millionth of a second.

It had been known since the nineteenth century that if an electrically charged particle is made to move in a circular path, it will constantly emit radiation. This implied that an electron that orbited a nucleus would rapidly radiate away all its energy of motion. It seemed that collisions between electrons and nuclei were inevitable. After all, orbital motion was the only thing that could counteract the attraction between the electrons and the nuclei. If the electrons were not revolving, there would be no centrifugal force and they would fall toward the nuclei even more rapidly.

In 1913, a solution to the problem was suggested by Niels

Bohr, a Danish physicist who had also spent a short time as one of Thomson's research assistants. Apparently he and J.J. did not get along particularly well. After a short stay at Thomson's laboratory, Bohr left and went to work under Rutherford instead.

Bohr was still working as Rutherford's assistant when he proposed his theory. He had no more idea than Rutherford how electrons might manage to move in circular orbits and yet not radiate. So he dealt with the problem by legislating it away. One of the assumptions upon which Bohr's theory was based was that electrons always radiated energy in this manner *except* when they were confined to atoms. But Bohr did not explain why this should be the case.

Such a theory would not even have been considered by Bohr's contemporaries if there had been nothing more to it than that. However, there was a great deal more. When Bohr worked out his theory in detail, he realized that it could be used to explain a large number of experimentally observed facts.

Bohr realized that an adequate atomic theory had to explain why atoms emitted and absorbed light of certain definite wavelengths. It had long been known that when an electrical discharge was passed through a gas, light was given off. When this light was passed through a prism, a continuous spectrum of colors was not observed. The light was found, instead, to be made up of a series of *spectral lines*, between which one could see vast dark spaces. Each line corresponded to a definite wavelength, the dark spaces to the wavelengths that were absent.

In order to explain the spectral lines, Bohr made the further assumption that only certain specific electron orbits were possible. An electron could move in orbit *A*, orbit *B*, and so on, but not anywhere in between. Light, Bohr said, was emitted when an electron jumped from one orbit to another, giving up energy in the process. If there were many different possible orbits, jumps between them could release different quantities of energy and produce rays of light that corresponded to the wavelengths that were seen in the laboratory. The dark spaces in the spectrum were explained by the fact that most orbits were impossible.

Many of Bohr's contemporaries found this idea to be quite bizarre. If Bohr was right, then it appeared that one had to accept the idea that an electron moving in a particular orbit could suddenly dematerialize and just as suddenly reappear somewhere else. The German physicists Otto Stern and Max von Laue found Bohr's ideas so shocking that they vowed to give up physics if, by chance, the theory turned out to be correct. Einstein, on the other hand, was delighted to hear of the new ideas. When Bohr's theory was reported to him, he commented that he had once had similar ideas himself but had not dared to pursue them.

Within a relatively short time, Bohr's *quantum theory* of the atom was provisionally accepted by the community of physicists. Not only did the theory allow physicists to make some sense out of the confusing mass of spectroscopic data that had been accumulated over the years, but it also seemed capable of explaining atomic structure and the chemical properties of different elements. Bohr's theory received experimental support when the German physicists James Franck and Gustav Hertz showed that when gases and vapors were bombarded with electrons, these electrons could absorb energy only in certain definite amounts. This result confirmed Bohr's assumption that only certain specific orbits were possible, and that atoms could absorb energy only in amounts (called *quanta*) that corresponded to jumps between orbits. Before long, it was apparent that this strange new theory was going to be successful. Stern and von Laue were apparently impressed. Not only did they renounce their vow to give up physics; they went on to make significant contributions to quantum theory themselves.

And yet the theory only served to make the electron seem even more mysterious an object than it had been before. No one could explain, for example, how it got from one orbit to another. As Rutherford pointed out, it appeared that an electron would have to "know" beforehand to which of the other possible orbits it was going to jump. Otherwise, it could not emit light of a single, definite wavelength when it began its leap.

Furthermore, the theory failed miserably under some circumstances. Although it explained why spectral lines should exist, and gave valuable insights into the problems of atomic structure, it did not yield the correct results when one tried to use it to calculate the wavelengths of the lines in the spectrum of the second-simplest atom, helium. Bohr's theory worked well only for hydrogen and for atoms of metals like sodium and potassium, in which a single outer electron orbited around the inner electrons and the nucleus. If one had to consider the behavior of more than one electron at a time, the theory gave results that were in conflict with experiment.

The difficulties remained unresolved until the mid-1920s, when a series of new discoveries about the electron were made. As physicists learned more about this tiny particle, they began to realize that it was not the minute, charged sphere that Thomson had envisioned. As their conception of the electron changed, their ideas about the nature of physical reality metamorphosed also. They began to realize that material objects were not made up of collections of tiny particles that behaved more or less like objects in the macroscopic world. Physical reality, which had once appeared to be an absurdly simple thing, was turning out to be a more complex affair than they had expected. They began to realize that the quantum world was not like the everyday world that they observed with their eyes. And, as they began to peer into this realm and try to understand the nature of quantum reality, they found that their investigations of the electron could provide a key to understanding.

Before I discuss the discoveries that were made by the quantum physicists of the 1920s, it will be necessary to go back a bit and to look at a discovery that was announced in 1900. In that year, the German physicist Max Planck propounded his quantum theory of radiation.

Planck had been pursuing some theoretical studies of *blackbody radiation,* a type of radiation that objects emit when they are heated. Unlike the radiation that is created by electrical discharges in gases (the kind that Bohr's theory explained), blackbody radiation is characterized by a *continuous spectrum* of wavelengths. For example, when a piece of metal is heated until it becomes white-hot, it emits light at all visible wavelengths, and in the ultraviolet and infrared regions of the spectrum as well.

Planck discovered that if blackbody radiation was to be adequately explained, it was necessary to make the assumption that atoms could emit or absorb energy only in discrete packets, or quanta. An atom in a hot object could give off one quantum of energy in a given period of time, or a hundred quanta, or a thousand. But it could not emit a fraction of a quantum of energy, or one and one-half quanta or five hundred thirty-six and two-thirds.

Planck found this result to be baffling. In fact, he spent years trying to find a way to avoid this conclusion. According to the accepted wave theory of light, such behavior was inexplicable. It was as though one were forced to come to the conclusion that an ocean wave could contain ten cubic feet of water, or twenty-seven cubic feet, but not sixteen and one-eighth, or nineteen and one-quarter.

Although Planck had reservations, Einstein accepted this result with alacrity. In 1905 (the same year in which he published his papers on Brownian movement and on relativity), he elaborated upon Planck's idea by suggesting that if light could be emitted or absorbed only in discrete packets of energy, one should conclude that light was made up of particles. Einstein was well aware that there existed a mass of experimental evidence which seemed to demonstrate that light was a wave phenomenon. But this did not trouble him; according to his theory, light had particle characteristics too.

Many of the physicists of the day considered Einstein's conclusion to be absurd. In their view, light had to be either one thing or the other; it could not be both at the same time. After all, in the everyday, macroscopic world, "wave" and "particle" were mutually exclusive categories.

One might object that an ocean wave is made up of particles. After all, water is composed of atoms and molecules. But this is only a quibble. Einstein was not saying that numerous small particles grouped themselves together into waves. In his theory, a particle of light was not a tiny constituent of a wave; he was thinking of light particles (called *photons* today) as objects that corresponded to whole series of wave crests and troughs.

When Bohr proposed his atomic theory in 1913, he drew on Planck's hypothesis. Planck's theory was a quantum theory of radiation; Bohr's was a quantum theory of the atom. The former implied that light could be emitted only in bundles of energy of a definite size; the latter explained why this was so. But when Bohr proposed his theory, he was still skeptical of Einstein's idea. Many other physicists went even further; they called it outrageous.

However, the hypothesis that light was made up of particles

eventually had to be accepted. When Einstein published his 1905 paper on the subject, he suggested certain experiments that could be performed to test his theory. When these, and other, experiments were done, the results confirmed that Einstein had been correct.

During the early 1920s, the French physicist Prince Louis de Broglie observed that, although light seemed to have a dual character, the accepted wisdom was that the electron did not. Physicists still thought of it as a material particle. De Broglie began to wonder if this conclusion was correct. Before long, he had set out to see if he could devise a theory of electron waves.

De Broglie presented his theory in his doctoral dissertation in 1924. His examiners didn't know what to make of it. So one of his professors showed de Broglie's thesis to Einstein, who commented that the idea might turn out to be an important discovery. De Broglie was given his degree, and shortly afterwards, Einstein published a paper in which he called attention to the hypothesis.

By this time, Einstein's prestige was enormous. But this did not prevent de Broglie's hypothesis from being greeted with derision in some quarters. One physicist went so far as to refer to it as *la Comédie Française*. But the skeptics again turned out to be wrong. De Broglie's theory was confirmed in 1927 when the English physicist George Thomson (J.J.'s son) and the American physicists Clinton Davisson and Lester Germer performed experiments which exhibited the electron's wave characteristics.

De Broglie's wave theory was only a beginning. Before it could be used to make accurate predictions of the behavior of atoms and of electrons, there was more work to be done. In the meantime, other physicists continued to elaborate upon Bohr's theory of the atom. In 1925, the Dutch physicists George Uhlenbeck and Samuel Goudsmit took the next step by proposing that electrons had spin.

Uhlenbeck and Goudsmit had observed that when light-emitting atoms were placed in a magnetic field, single spectral lines had a tendency to split into two lines, or three, or even more. A single wavelength of light was replaced by light that was emitted at a number of distinct wavelengths. Although the simplest cases could be explained by Bohr's theory, the more complicated ones could not.

Uhlenbeck and Goudsmit observed that if an electron did spin, it would behave as though it were a tiny magnet. It had been known for over a century that moving electrical charges produced magnetic fields. And of course spin was a kind of motion. If the electron did have magnetic characteristics, the two Dutch physicists reasoned, the interaction between the electron and an external magnetic field could cause the splitting. Depending upon the direction in which an electron was spinning, its energy would be raised or lowered. Since different wavelengths of light corresponded to different energy jumps, the spectral lines would split.

Uhlenbeck and Goudsmit, who were graduate students at the time, wrote a paper about their hypothesis and gave it to their professor, Paul Ehrenfest, asking him to send it off to a scientific journal if he thought it important. And then they began to have doubts. According to their calculations, the magnetic field of an electron was exactly twice as strong as it should have been, given the magnitude of the spin that they had computed. Other characteristics of the spin hypothesis seemed even stranger. For example, it appeared that the north pole of a spinning electron had either to align itself with the external magnetic field or to point in the opposite direction. If the theory was correct, the electron could not orient itself in any manner in between. Finally, a simple calculation showed that the perimeter of a spinning electron had to travel faster than the velocity of light.* But, according to Einstein's special theory of relativity, velocities greater than that of light were impossible.

So Uhlenbeck and Goudsmit went to their professor and told him that they had decided that their paper should not be submitted to a journal after all. When they did, Ehrenfest informed them that he had already sent it off for publication. They were young enough, he added, that making fools of themselves would not irreparably damage their reputations.

Of course, nothing of the sort happened. Uhlenbeck and Goudsmit made their reputations with their theory. Spin is a fundamental part of the concept of the electron today. Numerous

---

* At the time, the electron was thought to be a solid object with a definite diameter and circumference. As we shall see, this assumption was later shown to be inaccurate.

experiments have confirmed that an electron spin can have only two orientations (which physicists refer to as "up" and "down") when an atom is placed in a magnetic field, and the factor of 2 that so bothered Uhlenbeck and Goudsmit has been explained as a relativistic effect. Finally, no one worries about the velocity of the perimeter of the electron any more, because the electron is no longer thought of as a tiny sphere with a definite radius and circumference.

I will return to some of these points presently, and explain just how the electron is conceived today. At the moment, however, it might be best to continue with an account of the discoveries that were made during the 1920s. The next significant advance took place just a year after the publication of the Uhlenbeck-Goudsmit spin theory. In 1926, the Austrian physicist Erwin Schrödinger elaborated upon de Broglie's hypothesis and proposed a theory that was to become known as *wave mechanics*.

Schrödinger's theory gave physicists important insights into the nature of the electron and of the atom. In particular, it explained why only certain electron orbits existed. An orbit, Schrödinger pointed out, had to correspond to a wave with a whole number of crests and troughs; it had to be made up of one wave, or of two waves, or of eighteen or a hundred. Fractions were not permitted, because then the two ends of a circular wave pattern would not match up.

The theory also explained why electrons that were confined to atoms did not constantly radiate energy and spiral into the nucleus. The waves Schrödinger envisioned were *standing waves* that were analogous to the vibrations that are observed when a violin string is plucked. A violin string vibrates up and down; crests and troughs do not move from one end of the instrument to the other. In other words, Schrödinger's theory implied that an electron surrounded, rather than revolved around, the nucleus. If the electron was not really moving in a circle, then it would not have to be constantly radiating energy.

By the time that Schrödinger's theory was proposed, physicists had stopped laughing at every new idea. Too many bizarre conceptions had turned out to have experimental confirmation. But there was one thing that was very puzzling. During the previous year, 1925, the German physicist Werner Heisenberg had advanced a theory of his own, called *matrix mechanics*, which also explained the behavior of electrons in atoms. The two theories seemed to be based on entirely different sets of assumptions. Yet they gave the same results, results which could be verified by experiment.

Unlike Schrödinger's, Heisenberg's theory did not depend upon any intuitive picture of the behavior of electrons. It had an abstract mathematical character, and it was based on certain mathematical quantities called *matrices*. However, it is probably not inaccurate to state that while Schrödinger conceived of electrons as waves, Heisenberg's theory fit in with the conception of an electron as a particle.

Before long, it had been shown that although the two versions of *quantum mechanics* had few obvious resemblances to one another, they were really mathematically equivalent. But this left an important problem still unsolved. What, exactly, did Schrödinger's waves represent? Just what was it that was "waving"?

At first, Schrödinger thought that the waves might represent the density of electric charge. Perhaps the electron was not localized in any particular place, but spread out in a cloudlike pattern around the nucleus.

Although the idea of an electron cloud had certain attractions, it was soon rejected by physicists as untenable. An alternative interpretation, which had been proposed by the German physicist Max Born, was adopted in its place. Born showed that the only consistent way to interpret the theory was to conclude that Schrödinger's waves were a measure of the probability that the electron occupied any given position.

This probability-wave interpretation allowed the wave and particle descriptions of the electron to be combined. The electron was described as a wave pattern. Knowledge of the electron waves allowed one to compute the probability that the electron-as-particle would be found in any given place.

Acceptance of this interpretation had a very important consequence. It forced physicists to give up the deterministic outlook that had previously been a part of all physical theories. Scientists had always believed that if one knew what a particle was doing at any moment in time, and if one knew the forces that were acting upon it, it would be possible (at least in principle) to compute the particle's future behavior for all eternity. In quantum mechanics, this appeared not to be the case. The location of an electron in space was described by probabilities. Moreover, quantum mechanics did not allow one to compute exactly when an electron would jump from one orbit to another. It was possible only to compute the probability that this would happen during any given time period.

There is much that can be said about the probabilistic character of quantum mechanics. Indeed, the concept raises questions which have not been completely resolved today. Physicists began arguing about this, and related questions, shortly after quantum mechanics was discovered, and they are still arguing. I will return to this topic in another chapter, and discuss it at length. For the moment, however, it might be best to continue with the account that has just been interrupted.

Physicists no longer think of an electron as something which can orbit the nucleus of an atom. The concept of orbits has been discarded and replaced with that of *energy levels*. When an electron emits or absorbs a photon of light, it moves from one energy level to another. One set of wave probabilities is replaced by another.

If it is not accurate to think of an electron as an orbiting particle, and if it is not reasonable to view it as a cloud of electric charge, then there is probably no visual picture that is completely accurate. The best that one can do if one wants to describe the behavior of an electron in an atom is to say that it is located somewhere away from the nucleus. One cannot say exactly where, because quantum mechanics allows one to speak only of probabilities. One cannot think of an electron as something that resembles an ocean wave, because it is a wave and particle at the same time. The best that one can do is think of it as a thing with mass and charge that has no definite location within an atom.

It is not easy to visualize precisely what an electron "is," because the pictures that we form in our minds are dependent upon our experiences with everyday objects in the macroscopic world. We find it hard to conceive of something as being both a wave and a particle because the objects and phenomena we perceive with our eyes never exhibit both characteristics simultaneously.

However, it should not be considered surprising that electrons should have such odd properties, or that it should be so difficult to form mental pictures of their behavior. It would be more surprising if the objects that make up the microscopic world mimicked their macroscopic counterparts. After all, there is no reason to think that the world of the very small (or the very large, for that matter) should look like the world of everyday experience.

Although the quantum mechanics of Schrödinger and Heisenberg gives us some insight into the nature of the electron, it does not provide us with the final word on this subject. Quantum mechanics describes the behavior of electrons only in an approximate way. Electrons often move about at velocities approaching that of light (though electrons don't really move in orbits, they do have velocities; these velocities are also represented as probabilities). Since Einstein's special theory of relativity must be used if one is to have any hope of accurately describing the behavior of objects that move at high, *relativistic* velocities, a theory which amalgamates special relativity and quantum mechanics is needed if one is to obtain accurate results.

And then there is the matter of electron spin. As we have seen, the concept of spin was developed in the context of the Bohr theory of the atom. Although spin can be worked into the framework of nonrelativistic quantum mechanics, it is something that must be "added on." The idea of spin does not arise out of the theory in a natural way. As a result, nonrelativistic quantum mechanics tells us little about what spin "really is."

These two problems were solved simultaneously by the English physicist P. A. M. (Paul Adrien Maurice) Dirac in 1928. Dirac found a way to combine quantum mechanics with Einstein's theory, and to produce a theory which could account for certain discrepancies between the predictions of the nonrelativistic theory and experiment. Dirac's theory also provided a natural explanation of spin. Dirac found that the waves which described the behavior of electrons had two components, corresponding to the two possible spin orientations.

But Dirac's solution of the problem also raised new questions. The equations of relativistic quantum mechanics seemed to indicate that an electron could exist in two different kinds of energy state. Its energy could be either positive or negative. Furthermore, the negative energy states that the theory predicted were not so easy to interpret. If an electron that occupied one of these states was attracted by a positively charged particle, such as a proton, it would move away from the particle, not toward it. When one "pulled" on a negative-energy electron, it would respond by moving away; when one "pushed" on it, it would move nearer.

This kind of bizarre behavior had never been observed in nature, so it seemed that something must be wrong. One might think that Dirac could simply have disregarded the negative-energy solutions of his equations. After all, this is a procedure that scientists follow under other circumstances. For example, suppose that one is told that a square piece of plywood has an area of four square feet. It is immediately obvious that the piece must measure two feet on each side; the square root of 4 is 2. Now it so happens that the number 4 has two square roots, +2 and −2. Of course, 2 × 2 equals 4. But so does −2 × −2; when the multiplication is carried out, the minus signs cancel one another. But one does not say that a piece of wood can have sides which have lengths of minus two feet. One disregards the negative solution.

Unfortunately, this kind of procedure could not be carried out in the case of the Dirac electron. Quantum mechanics can be made consistent only if one deals with what is known mathematically as a *complete set* of solutions. Dirac's positive energy states did not form a complete set. Thus saying that electrons could have only positive energies would have led to contradictions.

Thus it seemed that one had to accept the idea that these bizarre negative energy states were real. Unfortunately, this led to conclusions which seemed to make the theory nonsensical. An electron will always seek the lowest possible energy level. Thus Dirac's theory seemed to imply that electrons would spontaneously jump from positive to negative energy states, giving up energy in the process. And since the number of possible negative energy levels was infinite, the electrons would go on jumping to lower and lower states until they possessed infinite negative energy.

The idea of "larger" amounts of negative energy is one that can easily seem a bit confusing. A good analogy is provided by the use of credit cards. Suppose one has a bank balance of $200, a positive amount of money. If one then charges $500 on a credit card, one is $300 in debt. If there were no credit limits, it would be possible to keep on making purchases until the debt approached infinity. Banks will not allow people to do this, but there is nothing to keep the Dirac electron from attaining a higher and higher "negative balance."

Dirac realized that there was only one thing that could prevent this kind of behavior from occurring. The fact that only positive-energy electrons were observed implied that all of the negative energy levels were already filled; there had to be a "sea" of an infinite number of negative-energy electrons.

Dirac's idea was based on a concept called the *Pauli principle*, or *exclusion principle*, which the Austrian physicist Wolfgang Pauli had proposed in 1925. According to the exclusion principle, no two electrons in an atom can occupy the same energy level. If the lowest energy level in a many-electron atom is occupied by an electron, then the other electrons can never jump down into this level unless the first electron is somehow displaced. The exclusion principle is confirmed by numerous experimental results. In fact, it is impossible to explain atomic structure without it.

Dirac realized that if a negative-energy "sea" existed, then positive-energy electrons would not be able to radiate away energy and drop into the negative states. Under ordinary circumstances, the negative-energy electrons would be unobservable. They could be detected only if one of them happened to absorb enough energy to move into the positive region. This would create an unoccupied level, or "hole," that could be seen.

In order to see what such a hole would look like, consider the following analogy. Suppose that a column of soldiers is standing at attention. Now suppose that one of the soldiers who is standing near the front of the column is ordered to fall out. This will leave an empty space in the column. If the other soldiers now step forward, one by one, in order to fill up the space, the "hole" will appear to move back until it reaches the rear of the column.

Dirac pointed out that a hole in a negative-energy sea would look like a particle with positive energy. The only difference between it and an electron would be that it would seem to have positive charge. Thus he suggested, in 1930, that the electron must have a positively charged counterpart. This particle would have the same mass as an electron, but would have a positive, rather than a negative, charge.

By 1930, physicists had grown accustomed to the fact that quantum mechanics forced them to accept certain bizarre conclusions about the behavior of subatomic particles. However, some of them found Dirac's theory a bit too much to swallow. If one accepted the theory, it was necessary to believe not only that electrons could exist in negative energy states, but also that these negative-energy electrons existed in infinite numbers. To make things worse, one was supposed to conclude that this sea of electrons contained holes which corresponded to particles that had never been observed.

However, this last state of affairs did not last for long. Just two years later, in 1932, the American physicist Carl Anderson discovered the positively charged electron, or *positron*, in cosmic rays. Dirac's theory had been vindicated.

The positron is said to be the electron's *antiparticle*. Today we know that subatomic particles always come in pairs. There are protons and there are antiprotons, neutrons and antineutrons, and so on. Antiparticles are ordinarily seen only in experiments conducted with modern high-energy particle accelerators. They are not often observed in nature, because particles and antiparticles annihilate one another when they meet, giving off a burst of radiation in the process (according to the picture proposed by Dirac, the particle falls into the antiparticle hole, and gives up energy as it does so). Since there are many more particles than antiparticles in nature, the latter do not survive for long once they are created.

It seems natural to ask how seriously one is supposed to take Dirac's idea that there is an infinite sea of negative-energy electrons. Is this sea real, or is it only a useful fiction? This is not an easy question to answer. If a sea of negative-energy electrons really exists, there is no way to detect it experimentally. One can observe only the "holes" which correspond to the electron's antiparticle, the positron.

Dirac's idea that there can be holes in a negative-energy sea is not the only interpretation possible. During the 1940s, the Swiss physicist Ernst Stückelberg and the American physicist Richard

Feynman showed independently that there is another way in which the existence of positrons (and other antiparticles) can be interpreted. They can be regarded as electrons that travel backwards in time. According to this interpretation, a positron is not a hole at all; it is a time-reversed electron.

There is really no way of telling whether a sea of negative-energy electrons exists, or whether electrons sometimes travel backwards in time, or whether neither of these interpretations is correct. The problem is that there is no way to distinguish between the different interpretations; in each case, one obtains the same theoretical results.

Perhaps it is simplest to think of particles and antiparticles as objects which are, in some sense, mirror images of one another. Indeed, this is the view that is normally adopted by physicists who work with *quantum electrodynamics*, a refinement of Dirac's theory that will be discussed in the next chapter.

But what are electrons and positrons really like? No one really knows. It is not even obvious that the question is particularly meaningful. What physicists demand of a theory is that it be capable of producing numerical predictions that can be tested experimentally. If these predictions are verified within the limits of some reasonable margin of error, we say that the theory has been confirmed. However, the task of interpreting a theory, of forming a mental picture of the kind of reality that the theory implies, is not part of the verification process.

Trying to determine what a theory "means" can be very difficult. It is safe to say that no one can really visualize what an electron "looks like," even though theory can successfully predict the results that will be obtained when experiments with electrons are performed. For that matter, there is not yet any universal agreement about the meaning of quantum mechanics, even though the theory was developed as long ago as the 1920s. The discovery of quantum mechanics produced questions about the nature of reality that have not yet been completely answered.

I will return to this topic from time to time in the chapters that follow, and discuss these questions in somewhat more detail. In fact, as will soon become apparent, questions about the meaning of physical theories constitute the real subject matter of this book. At the moment, however, there are a few more things to be said about the electron, and about the validity of Dirac's relativistic theory.

From time to time, I have made reference to the phenomenon of electron spin, and have pointed out that a spinning electron does not behave quite like an ordinary rotating object. It is not possible to visualize the electron as a tiny spinning sphere. Although there is some resemblance, one should not make the mistake of assuming that the resemblance is too exact.

The best way to describe the difference is by means of an example. Suppose that the earth were somehow tilted through an angle of 180 degrees. If this were done, the positions of the north and south poles would be reversed. An additional rotation of 180 degrees would then restore the two poles to their former positions. A 360-degree rotation, in other words, would leave the earth just as it was before.

This is not true in the case of an electron. A rotation of 720 (twice 360) degrees is required to bring an electron back to its original orientation. The "poles" of the electron have to be rotated through a complete circle twice. Incidentally, this fact is related to Uhlenbeck and Goudsmit's discovery that the magnetic field produced by an electron was exactly twice what it should have been.

This sounds somewhat paradoxical. However, this is the case only if we insist on regarding an electron as a tiny sphere of the sort that Thomson envisioned. If we accept the idea that it is not, the "paradox" disappears. A top, a baseball, or a spinning planet could not behave in this manner. An electron can because it is a different kind of object.

However, we have not yet reached the point where we can say precisely what kind of object it is, for Dirac's theory does not provide the final word about the electron. There are small discrepancies between its predictions and experimental results.

The reason that these discrepancies exist is that Dirac's theory does not take the interaction between particles and fields into account. Electrons, for example, interact with the electromagnetic fields associated with light and with other forms of radiation. Dirac's theory represented a great step forward. However, it describes the electron as an inert, noninteracting object. Consequently the picture that it creates of electron behavior cannot be completely correct.

The first steps toward the development of theories that could describe interactions between particles and fields were taken during the late 1920s. As one might expect, the development of these field theories led to further changes in physicists' conceptions of the nature of the electron and of subatomic reality.

Before these theories are discussed, it might not be a bad idea to discuss the idea of force, and the changes that have taken place in physicists' idea about what a force is. After all, the notions of field and force are intimately related. It was through the idea of "fields of force" that the field concept was first introduced into physics.

2

Forces and Fields

Shortly after Isaac Newton had proposed his law of gravitation in 1687, he was criticized by the German philosopher and mathematician Gottfried Wilhelm von Leibniz. Leibniz objected to the idea that gravitational forces could act across empty space. If Newton had been able to explain how gravitational force was transmitted, he said, he might take the theory more seriously. The idea of action at a distance, however, was unacceptable. Such a concept made gravity a "perpetual miracle."

Leibniz' misgivings were widely shared. To many of Newton's contemporaries, acceptance of the idea that celestial bodies could exert forces on one another over great distances was tantamount to reviving the idea of "occult properties" from the discredited Aristotelian physics. Forces, the critics objected, could only be transmitted between bodies that were in direct contact.

When Newton's critics made this objection, they were not accusing him of ascribing a supernatural character to gravity. The original meaning of the word "occult" is "hidden." In Aristotelian physics, occult properties were such things as sympathies and antipathies. One seventeenth-century account described them as "friendly affections (or their opposites), or coordinations and innate relations of one thing to another."

Such criticisms put Newton on the defensive. He replied that he had intended only to describe the law of gravity in a mathematical way, and that he did not pretend to know its cause or mechanism. "I make no hypotheses," he said on one occasion. On another, in a letter to Richard Bentley, the bishop of Worcester, he elaborated upon this:

> That gravity should be innate, inherent, and essential to matter, so that one body may act upon another at a distance through a *vacuum*, without the mediation of anything else, by and through which their action and force may be conveyed from one to another, is to me so great an absurdity that I believe no man who has in philosophical matters a competent facility for thinking can ever fall into it. Gravity must be caused by an agent acting constantly according to certain laws, but whether this agent be material or immaterial I have left to the consideration of my readers.

Apparently the problem continued to trouble Newton, for, in the end, he did make hypotheses concerning the nature of gravitational force. In the second edition of his *Principia Mathematica*, the book in which his theory of gravity was explained, Newton suggested that gravitational forces might be transmitted by the *ether*, a tenuous form of matter which supposedly filled all space, including the spaces between the sun and the planets.

The idea of an ether was not original with Newton. The concept had originally been proposed by Aristotle, who believed that a perfect vacuum was impossible. In Aristotle's opinion, terrestrial objects were composed of the four elements, earth, air, fire and water, while the celestial realm was made up of a fifth element, the ether, that was in constant, circular motion.

But Newton's suggestion that gravity might be transmitted by the ether did not really solve the difficulty. According to Newton, this hypothetical substance was presumably composed of tiny particles. If one invoked their existence in order to explain gravity, the problem of action at a distance would not really be solved, for one would now have to explain how the ether particles exerted forces on one another. Interatomic action at a distance is no easier to explain than the gravitational variety; the only difference is that the distances between bodies are smaller.

Newton did not work out his hypothesis of a corpuscular ether in detail, and there is no way of knowing whether he really believed in the idea or not. Newton did think that an ether existed; he invoked its presence to explain other physical phenomena, such as the reflection and refraction of light. However, when he suggested that it might be made up of discrete particles, he may have been doing no more than throwing out an idea which he thought might make his theory of universal gravitation a bit more plausible.

As Max Planck once pointed out, revolutionary new ideas in physics do not always gain acceptance because their opponents eventually become convinced of their truth. They are often accepted because their opponents eventually grow old and die. Such was the case with Newton's law of gravitation. Those who viewed gravity as an occult quality died, and a new generation of scientists grew up with Newton's theory, became familiar with it, and elaborated upon it. Although they did not know what gravity was either, Newton's theory was too useful to give up. Not only did it explain the motion of planetary bodies, it also accounted for the behavior of such objects as comets and explained the phenomenon of ocean tides.

One of the scientists who elaborated upon Newton's ideas was the eighteenth-century French mathematician and astronomer Joseph Louis Lagrange. Lagrange worked out mathematical methods for dealing with problems that Newton had been unable to solve, such as the calculation of the orbit of a planet whose motion is perturbed by the gravitational attraction of other planetary bodies (the orbital motion of Mars, for example, is influenced by the other planets, most significantly by Jupiter).

In the course of solving some of the problems of celestial mechanics,* Lagrange introduced an idea which was to be quite

---

* Mechanics is the branch of physics that deals with the motion of bodies. Hence celestial mechanics is the discipline that deals with the motion of celestial bodies.

influential upon subsequent thought concerning the nature of action at a distance. Lagrange saw that it was possible to describe gravitation mathematically by defining a gravitational *potential* at every point in space. Potential was a quantity which allowed one to deduce the magnitude and direction of the gravitational force that would be experienced by a small body introduced at any point in space.

Lagrange thought of gravitation not as a force which acted between two bodies, but as a field which existed everywhere. For example, if one could calculate the gravitational field produced by the sun, then it would be possible to calculate the forces acting on any body (such as the earth, or Saturn, or a comet) that moved within that field. Potential, in other words, was a set of numbers that described gravitational field intensity at every point in space.

It was not Lagrange's intention to replace the concept of action at a distance with that of a field of force. In fact, Lagrange never even used the word "field." His use of the concept of gravitational potential was nothing more than a simplifying mathematical device which made some of the more difficult problems in celestial mechanics more tractable. Nor did the concept of field explain how forces were transmitted over astronomical distances. The ideas of potential and of gravitational attraction were really mathematically equivalent, and nothing new was added when Lagrange introduced the former.

Nevertheless, Lagrange's approach was to lead to a revolution in physics. When his successors applied the idea of potential to electric and magnetic fields, new insights were gained and new discoveries did result. These discoveries led, in turn, to changes in physicists' conceptions of the nature of force.

The revolution began in 1820, when the Danish physicist Hans Christian Oersted announced a new experimental discovery. Oersted had found that when he caused an electric current to flow in a straight wire, a compass needle that was placed near the wire was deflected in a direction perpendicular to that of the flow of the current. Furthermore, when the electric current was reversed, the compass needle turned around and pointed in the opposite direction.

Oersted's discovery established, for the first time, that there was a connection between electricity and magnetism. The relationship was not a simple one, however, and it could not be explained in terms of attraction or repulsion. Stationary electric charges did not deflect a compass needle. The moving charges that constituted an electric current did not cause the needle to align itself in the direction that the charges were moving, but rather at right angles to their flow. It appeared that the only way to explain Oersted's discovery was to postulate that moving charges created a magnetic field, and that the field acted, in turn, on the magnetized compass needle.

Although Oersted's discovery was important, he did little to follow up on it. The English physicist Michael Faraday was thus the first to study the new phenomenon in detail. Faraday suspected that a magnetic field was nothing more than an electric field in motion. If this was true, he reasoned, then it followed that electric and magnetic forces could be converted into one another. If the moving charges that made up a current could deflect a magnet, then it ought to be possible to observe the reverse effect. Faraday carried out the experiment that this conclusion suggested, and his hypothesis was confirmed. An electric current was created when he moved a wire about in the magnetic field of a magnet.

These discoveries had immense practical significance. Faraday made use of them to construct electric motors and transformers, and the first electrical generator. An electric motor is nothing more than a device in which electric currents induce motion in magnets. In a generator, moving magnets produce the reverse effect; electric currents are the result. There is a famous, but possibly apocryphal, story about Faraday's discovery of the latter effect. After observing a demonstration in which Faraday generated electricity in this manner, William Gladstone (who was later to become a noted British prime minister) is supposed to have asked, "But, Mr. Faraday, what is the use of this?" Faraday is supposed to have responded, "Sir, in twenty years, you will be taxing it."

The theoretical implications of Faraday's work were equally important. His experiments prompted speculation about the nature of the forces transmitted by magnets and by charged particles. Although these speculations led to a concept of force that was really not very accurate, they nevertheless constituted a significant advance. The development of erroneous ideas in science does not always lead to disaster. It is sometimes necessary to form mistaken ideas before it is possible to grope one's way toward the truth. A mistake, after all, can be repudiated when experimental evidence shows that it is based on hypotheses that are untenable. Forming incorrect conclusions about the nature of a phenomenon is better than having no ideas at all. In the latter case, one would not know what theoretical investigations should be undertaken or what experiments should be performed.

Scientists have never claimed to "have all the answers." There is a popular myth which suggests that they are often guilty of just this kind of arrogance. But this myth is based on a misunderstanding of the nature of science. The true subject matter of science is not that which is known, but rather the unknown. Scientists are men and women who spend their lives attempting to solve puzzles. Those who are good scientists know that they will not find any final answers. They realize that most solutions are tentative, that it is the business of scientists to discover what is wrong with existing solutions, and to propose better ones.

I think, therefore, that it will be worthwhile to discuss the speculations that Faraday's work elicited, even though many of the ideas upon which these speculations were based were later shown to be false. Too many accounts of scientific activity focus only on the successes. In doing so, they paint an inaccurate picture of the nature of scientific discovery.

When Faraday's contemporaries speculated about the nature of electric and magnetic fields, they generally made use of Newton's idea of an ether that filled all of space. It was believed that fields—electric and magnetic fields, and gravitational fields as well—were the result of tensions or strains in this ether. For example, it was presumed that a gravitating body produced stresses in the ether around it, and that these stresses caused the ether to transmit the forces which acted on other massive bodies. Electric and magnetic forces were thought to be transmitted in a similar manner. The only difference was that the strains that they caused were of a different type.

It was perhaps the English physicist William Thomson—better known today as Lord Kelvin (he was raised to the peerage in recognition of his scientific work near the end of his life)—who

elaborated upon such ideas in the greatest detail. According to
Thomson, the ether was not composed of discrete particles, as
Newton had suggested. In his view, it was a kind of fluid. As has
been noted previously, the hypothesis of a corpuscular ether had
one very notable defect. It was not really capable of dealing with
the problem of action at a distance if one did not know how to
explain the nature of the forces between the ether particles. On
the other hand, if one assumes, as Thomson did, that the ether is
a continuous fluid, the problem disappears. If there are no ether
particles, one does not have to worry about the nature of "interatomic" forces.

In 1867, Thomson elaborated upon his ideas by publishing a
theory of "vortex atoms." The German physiologist and physicist
Hermann Helmholtz had previously published a theoretical paper in which he showed that a vortex in a perfect fluid would be
indestructible. Thomson suggested, therefore, that atoms might
be tiny vortices in the ether. According to his theory, the ether
was a space-filling plenum. The ultimate constituents of matter
were vortices, and vortex filaments were responsible for the trans-
mission of forces.

The next step in the development of ether theory took place
in 1873, when the great Scottish physicist James Clerk Maxwell
published his *Treatise on Electricity and Magnetism*. In this book,
Maxwell proposed a theory which not only explained all known
electrical and magnetic phenomena, but which also predicted
something quite new. According to Maxwell, if a changing electric
field could produce a magnetic field, and if a time-varying magnetic field could create an electric field also, then it followed that
electromagnetic waves should exist. Such waves, Maxwell said,
could theoretically be produced by oscillating electric charges.
Once the waves were created, they would propagate through the
ether at a high velocity.

This velocity, Maxwell pointed out, could be calculated from
certain electrical measurements that could be performed in the
laboratory. When Maxwell carried out this calculation, he found
that this velocity was exactly equal to that of light. He suggested,
therefore, that light was electromagnetic radiation. A ray of light
was made up of coupled electric and magnetic fields which oscil-
lated in directions at right angles to the direction of travel. In

other words, light was nothing more than *transverse* vibrations in the ether.

Physicists distinguish between two different types of wave motion. An example of the other kind, *longitudinal* vibrations, is provided by the phenomenon of sound. When sound waves are propagated through the air, the air molecules vibrate back and forth in a direction parallel to that in which the sound travels.

In Maxwell's time it was already known that although longitudinal waves could be set up in any kind of substance, transverse vibrations were transmitted only by solids. One was thus led to the rather paradoxical conclusion that the ether had to be a substance simultaneously rigid enough to transmit the transverse waves of light, and yet tenuous enough to allow the earth and other planetary bodies to pass through it unhindered.

One might think that this result would have caused physicists to question the existence of the ether. But this is not what happened. In fact, when the German physicist Heinrich Hertz succeeded in generating radio waves in 1887, and in showing that they possessed all the properties that Maxwell had ascribed to electromagnetic radiation, this was generally considered to be a confirmation of the ether's existence.

This created a situation which, from our point of view, was somewhat paradoxical. Maxwell's theory was a correct one in the sense that it could be verified by experiment. And yet it was based upon a model of reality that was erroneous, one that postulated the existence of an all-pervading ether. If such a thing is possible, then perhaps it is not so surprising that it is often difficult to tell how seriously one should take the picture of reality that a theory seems to imply.

Physicists still use Maxwell's theory today. However, the idea of an ether, upon which the theory once seemed to depend, was discarded long ago. Nowadays one speaks of oscillations of electric and magnetic fields in empty space; it is no longer thought necessary that there exist a medium for the fields to oscillate in.

But of course the physicists of the nineteenth century viewed matters differently. By the 1890s, many of them believed that ether theory had enabled them to successfully solve the problems of the nature of matter and the nature of forces between atoms (although there were a few doubters such as Mach and Ostwald, most scientists accepted the existence of atoms by this time). The physicists of the nineties believed that matter was composed either of ether vortices (as William Thomson suggested) or of tiny charged particles. Forces were strains in the ether. The third component of physical reality—radiation—was nothing more than ether vibrations. Since, by this time, it was known that heat was a phenomenon associated with molecular vibrations, there was really nothing else that might be included in the catalog. The universe was constructed of ether, electrical charges, and strains in the ether. There was nothing else.

To be sure, it was necessary that the ether possess some very odd properties. It apparently behaved like a tenuous gas and like a rigid solid at the same time. But perhaps that was really not as surprising as it seemed. After all, one would not expect it to resemble the more familiar forms of matter.

There was only one little flaw in this picture. Physicists were unable to discover a way that the ether could be directly observed. In fact, a crucial experiment had been performed by the American physicists Albert Michelson and Edward Morley in 1887, and had yielded a negative result.

Michelson and Morley had reasoned that, since the earth was moving around the sun at a velocity of about thirty kilometers per second, it should be possible to detect an ether "wind" blowing past the earth in the opposite direction. Michelson and Morley therefore performed an accurate experiment which was designed to measure the differences in the velocities of light rays that traveled in different directions with respect to the earth's motion. When light was traveling "downstream" with respect to the ether flow, it should travel faster than it did when it was directed into the wind.

Michelson and Morley did detect what seemed to be a slight difference in velocities. However, the result they obtained was much smaller than that which they expected. Since the difference they observed could easily be attributed to small experimental inaccuracies, they had to conclude that their attempt to measure the effects created by the ether wind had failed.

But no one suggested that the concept of the ether should be discarded. Physicists were loathe to take that course. The assumption that the ether existed seemed to explain numerous phenomena; it was not a hypothesis that could be given up lightly. Consequently physicists began to invent hypotheses that would explain the Michelson-Morley result while retaining the idea of an ether. Some suggested that no ether wind had been detected because the earth dragged the ether along with it as it revolved around the sun. The Dutch physicist Hendrik Lorentz and the Irish physicist George Francis FitzGerald independently advanced a theory that moving objects contracted in the direction of their motion. In their view, an ether wind had been blowing past the earth, but the contraction in Michelson and Morley's apparatus had affected their velocity measurements.*

Scientists are often reluctant to give up their theories, especially when those theories seem to possess great powers of explanation. When they are confronted with experimental evidence that raises questions about the validity of their theories, they often try to find ways to patch up the theoretical structure before they begin to consider other alternatives. Sometimes this method works; there are many theories that have undergone modification from time to time. But sometimes so many ad hoc hypotheses are added to a theory that it begins to look too contrived to be a reasonable explanation.

Scientists generally demand that a good theory should possess a certain quality of elegance, that it should explain a wide range of phenomena with a minimum of assumptions. History demonstrates that they are generally right to make this demand. It is the parsimonious, elegant theories that have most often turned out to be correct. On the other hand, theories that were made up of patchworks of diverse assumptions have generally turned out to be wrong.

But sometimes a patchwork theory can be retained for years. There are a number of reasons for this. For one thing, scientists who have built their professional lives around the elaboration of a theoretical idea are understandably reluctant to give it up. Sometimes genius of the highest order is required to see just what the valid alternatives are; a theory is often retained simply because no one has any better ideas.

* There is a limerick about a fencer named Fisk who moved so fast that his rapier flattened into a disk. The contraction that Lorentz and FitzGerald were postulating here was somewhat less extreme, however.

So perhaps we should not be surprised to discover that the ether theory lingered on during the first decade of the twentieth century, and that it took an Einstein to point out the obvious: the reason that no one had been able to detect the ether was that this substance simply didn't exist.

The ether concept was banished by the special theory of relativity, which was published in 1905, the same year in which Einstein propounded his theory of the Brownian movement and his quantum hypothesis. This was probably the only time in history that a single scientist presented three separate discoveries of such magnitude to the world in a single year. Any one of them would have been sufficient to make the name "Einstein" a notable one in the history of physics. That he was able to produce three such discoveries in the same year is close to miraculous.

Popular accounts of the discovery of special relativity sometimes give the impression that the theory was Einstein's response to the Michelson-Morley experiment. This is rather misleading. It is not even certain that Einstein knew of this experiment at the time that he put forward his relativity theory; he gave contradictory answers when asked about this during the latter part of his life. However, he did make it clear that the Michelson-Morley result played no part in his thinking during the period in which his theory was formulated. The idea of relativity was based on quite different considerations.

In 1895, when Einstein was sixteen years old, he began to wonder what would happen if one could travel at the velocity of light and observe a light ray that was moving in the same direction. He realized that the oscillating electric and magnetic fields that constitute a ray of light would appear to the observer to be frozen. Einstein wondered how this could be; such frozen fields were never observed in nature.

Years later, Einstein began to think about an old idea known as relativity. It had long been known that, as far as the laws of mechanics were concerned, relative motion was the only thing that counted. For example, Galileo had pointed out during the sixteenth century that if one dropped an object from the mast of a ship, it would fall straight down to the deck, no matter how fast or how slowly the ship was moving with respect to the ocean. The only thing that mattered was the motion of the falling object relative to the ship. The same phenomenon, incidentally, is commonly observed today. When a liquid is poured into a glass on an airplane, it flows in the same way that it would on the surface of the earth, even though the plane may be moving at a velocity of six hundred miles per hour. The laws of motion within the plane are the same as they are in a motionless building on the earth's surface.

Einstein observed that Maxwell's laws of electricity and magnetism did not seem to have this character. Maxwell's theory implied that a moving magnet would set up an electric current in a nearby wire. It implied, also, that if the magnet was held stationary and the wire was moved, the same current would be set up. According to Einstein's view, this was as it should be. It shouldn't make any difference whether it was the magnet or the wire that was moving; only relative motion mattered.

However, Maxwell's theory described these two phenomena in different ways. When the magnet was moved, an electric field was set up. It was this field which created the current in the wire. However, when the magnet was motionless, no such field was created, even though the same current was induced.

Maxwell's theory seemed to distinguish between a moving magnet and a moving wire. In one case, there was an electric field; in the other, there was not. It appeared that the laws of electricity and magnetism, unlike those of mechanics, distinguished between absolute and relative motion.

Einstein found it difficult to accept the idea that this distinction could be valid. So he set out to see what would happen if he made the assumption that the principle of relativity should apply to Maxwell's theory also, and attempted to see how the laws of electricity and magnetism could be modified in order to eliminate the distinction between absolute and relative motion.

Einstein found that if he wanted to obtain the desired result, it was necessary to assume that the velocity of light, as measured by any observer, was always constant. In other words, it was necessary to postulate that the speed of light would always be 300,000 kilometers (186,000 miles) per second, whether one was moving toward a source of light or away from it. Similarly, the same result would be obtained if the light source was in motion with respect to the observer.

At the time, this seemed to be quite an audacious assumption. After all, the velocities of ordinary objects do not exhibit this kind of behavior. If a jet plane is traveling at a speed of a thousand kilometers per hour, and if it fires a rocket that has a velocity of 800, the rocket will streak toward its target at a speed of 1800 kilometers per hour; the two velocities must be added. Yet, according to Einstein, light did not behave in this manner at all. If one attempted to add to the velocity of light by causing it to be emitted from a moving source, it would still move at a speed of 300,000 kilometers per second.

In his paper on relativity, which was published in the German journal *Annalen der Physik*, Einstein suggested that this made the concept of an ether "superfluous." If the velocity of light, as measured by any observer, was always constant, and if only relative motion could be detected, then it was not reasonable to believe that there existed a fixed ether. The presence of an ether implied a fixed frame of reference and the existence of absolute motion. But if there was no such thing as detectable absolute motion, then one had no need of the concept of an ether.

Einstein did much more than imply that this old—and by 1905, outmoded—idea should be discarded. His special theory of relativity successfully describes the behavior of objects which travel at velocities that are a significant fraction of the speed of light. Einstein found that when such velocities were attained, significant new phenomena would appear. For example, he found that rapidly moving objects would contract in the direction of their motion, just as Lorentz and FitzGerald had suggested. But Einstein did not view the phenomenon in quite the same way that his predecessors did. Einstein showed that the contraction was not an absolute effect; it too was relative.

This fact can best be explained by means of an example. Suppose that a space vehicle is traveling either away from or toward the earth at a velocity approaching that of light. To observers on the earth, the ship will appear to contract in the direction of its motion. It doesn't make any difference which way the ship is moving. The only thing that matters is the magnitude of the velocity. If the ship is moving at a velocity of 87 percent of the speed of light with respect to the earth, it will appear to shrink to one-half of its former length. However, since the effect is relative, observers on the ship will view matters differently. From their point of view, the ship will have the same length as before, but the earth will seem to flatten out.

Einstein's theory also predicted a *time dilation* effect. From the point of view of an earthbound observer, time on the ship will appear to pass more slowly. If a clock on the earth shows an elapsed time of one hour, only thirty minutes will seem to pass on the ship. Naturally, this effect is relative too. To the observers on the ship, it is the earthbound clocks that will seem to move more slowly.

The time dilation effect is often used by science fiction authors who imagine situations in which space vessels spend a relatively short time traveling through space. When these ships return to earth, centuries have passed. One might think that if time dilation is relative, such a thing could not happen. After all, one might object, wouldn't one expect the opposite effect if time were measured from the point of view of earthly observers? Although it is true that time will pass more slowly on the ship from the standpoint of an earth inhabitant, there is really no paradox. This is an example of a situation where the passage of time is not entirely relative. The time dilation effect is symmetrical only when different observers move at constant velocities with respect to one another. But this is clearly not the case here. If the ship is ever to return to the earth, it will have to, at some point, decelerate from its initial high velocity, come to a stop, and accelerate in the opposite direction in order to make the return journey. It is not legitimate to assume that, from the standpoint of the shipbound observers, the entire universe slows down, comes to a stop, and then accelerates in the opposite direction.

When one calculates the effects of time dilation under such circumstances, one finds that a much shorter length of time would indeed pass on a spaceship making a round trip to another star. Some day there may indeed be space travelers who spend a few years voyaging through space, and who discover that centuries have passed when they return to earth.

The time dilation effect has been verified experimentally. In 1976, an airplane that was carrying an accurate atomic clock was flown around a long racetrack. Even though the plane flew at a relatively slow speed (compared to that of light), the clock proved to be accurate enough to make the time dilation effect measurable. After the plane had circled the racetrack a number of times, it brought its atomic clock back to earth, and it was found that the elapsed time that it had measured was a small fraction of a second less than that measured by a clock that had remained on the ground.

According to Einstein's theory, there is yet another important effect which is observed when velocities approaching that of light are attained. This is the *relativistic mass increase*. When an object moves rapidly, its mass will appear to become greater. Again, the effect is symmetrical. From the viewpoint of observers on earth, the mass of an accelerating spaceship would increase without limit as it approached the velocity of light. Similarly, from the standpoint of observers on the ship, the mass of the earth—and of every object on its surface—would appear to increase in the same manner.

Like time dilation, the relativistic mass increase is quite real, not just an illusion. If a small dust particle that was traveling at a velocity very close to that of light were to strike the earth, the impact would be sufficient to cause our entire planet to burst into fragments.* The mass increase, incidentally, is observed experimentally every time a subatomic particle is accelerated to high velocities in the laboratory.

The most famous consequence of Einstein's special theory of relativity, the equivalence of mass and energy, is not to be found in his original paper on the theory. Einstein overlooked this result when he first worked out the theory. He soon remedied this defect, however, by discussing the equation $E = mc^2$ in a short paper that he sent to the same German scientific journal later the same year.

It is really not very difficult to see why mass and energy must be equivalent. One only has to consider the fact that an expenditure of energy is required if an object is to be accelerated to a high velocity. If this causes the body's mass to increase, it seems reasonable to surmise that there must be a connection between the en-

---

* One needn't worry that this will actually happen, however. There are no known naturally occurring phenomena that could cause a dust particle to be accelerated to such a velocity.

ergy that is "put in" and the mass that "comes out." In fact, when one analyzes the phenomenon in detail, the equation $E = mc^2$ immediately results. Then, if mass and energy are equivalent under these circumstances, there is every reason to think that the equivalence is valid in general. It would surely be odd if mass and energy could be converted into one another only when objects were accelerated to velocities near that of light.

The equivalence of mass and energy has, of course, been confirmed on numerous occasions. Mass is converted into energy every time a nuclear weapon is exploded, and it is converted into energy in every operating nuclear reactor. Energy is converted into mass whenever pairs of particles and antiparticles are created in cosmic ray showers, or in the laboratory. The energy that binds protons and neutrons together in atomic nuclei has a mass equivalent, and this mass has been measured. It has been known for decades that the mass of a nucleus is not equal to the masses of its constituent particles. Einstein's equation $E = mc^2$ is one of the best confirmed in modern physics.

The equivalence of mass and energy makes it possible to divide the mass of any moving object into two parts: its *rest mass*, and the mass that results from the relativistic mass increase. The concept of rest mass is important in the fields of physics that deal with subatomic particles. When one speaks of the "mass" of an electron or a proton, for example, it is the rest mass that is meant. The mass of such a particle will increase if it is accelerated to a high velocity. Hence it is important to be able to speak of the mass that such a particle would have if it were motionless. Naturally, the concept of rest mass can also be applied to macroscopic objects. However, it is not very important in this context, since we do not yet have the capability of accelerating large objects to relativistic velocities.

There is one other consequence of the mass increase that might be mentioned. According to Einstein's theory, the mass of any object would become infinite if it were accelerated to the speed of light. Since this would require the expenditure of an infinite amount of energy, one must conclude that material objects can never attain light velocity. If more and more energy were expended to accelerate them, they could approach the speed of light more and more closely. However, they could never quite reach it.

There seem to be more popular myths associated with special relativity than with any other theory in physics. According to one of them, special relativity is an especially difficult theory to understand. However, nothing could be farther from the truth. The theory is mathematically quite simple. Although the effects that it predicts in objects traveling at high velocities may seem bizarre to the person who has never encountered such ideas before, these effects are not difficult to comprehend.

It is true that when the theory was first published, many physicists failed to understand it. Most likely, the reason for this was not that the theory was too complicated, but that it was too simple. The physicists of the day had little trouble following Einstein's mathematical derivations. However, many of them just didn't get the point; they failed to understand what Einstein was getting at. Even Max Planck, who was one of the early supporters of Einstein's ideas, referred to special relativity, on occasion, as the "Einstein-Lorentz theory." He apparently didn't see, at first, that special relativity differed in fundamental ways from Lorentz' theories about the interactions between moving bodies and the ether. Lorentz had also deduced that such effects as a time dilation should exist. However, he thought that, like the length contraction he had postulated, it was an absolute, not a relative, effect. In confusing Einstein's theory with Lorentz', Planck remained oblivious to the fact that the fundamental idea of the special theory of relativity was relativity!

In some quarters, Einstein's theory was simply ignored. For example, some British physicists continued to propound theories about the ether even after Einstein had shown that the concept was unnecessary. In 1907, the British physicist Oliver Lodge read a paper on the motion of the ether at a meeting of the British Association for the Advancement of Science. According to Lodge, the ether had a density of thousands of tons per cubic millimeter and every part of it was "squirming with the velocity of light." In 1908, the noted French mathematician Jules Poincaré, who had also published ideas about relativity, and who is generally considered to be one of the forerunners of Einstein, proclaimed that "beyond the electrons and the ether there is nothing." In 1909, J. J. Thomson spoke of an "invisible universe" of ether.

Nevertheless, during the decade that followed the year 1905, Einstein's theory gradually won acceptance. Although there were no crucial experiments which forced acceptance of Einstein's ideas, physicists began to realize that the idea of an ether had become too cumbersome to be taken seriously. The more the concept of an ether was investigated theoretically, the more contradictory became the properties that the substance had to have. Special relativity was finally accepted because it provided a simple description of phenomena associated with rapidly moving objects. Its simplicity and transparent logic made it preferable.

The acceptance of Einstein's ideas created a new problem, however. Fields could not be interpreted as deformations of the ether if the ether did not exist. As a result, physicists found themselves confronted by the same problem that had so troubled Newton. They had no way of explaining action at a distance.

However, they continued to elaborate upon the field concept, even though they did not yet know what a field was. It is not always necessary to explain a phenomenon before one deals with it mathematically. Scientists found that if they continued to assume that various different kinds of fields existed, they could still work out theories and make mathematical predictions that could be confirmed by experiment. In the view of many physicists, this was all that one could reasonably demand.

Although Einstein's special theory of relativity did away with the ether, its acceptance did not cause physicists to have doubts about the existence of fields. On the contrary, it made the use of the field concept essential. The reason that it did this is that there is no place for the idea of instantaneous action at a distance in relativity.

The key word here is *instantaneous*. Special relativity not only implies that no material object can ever attain light velocity, it also tells us that no signal or causal influence can propagate at a speed faster than that of light. If such a signal did travel at greater than light velocity, it would seem to some observers to be traveling backwards in time; from their point of view, it would arrive at its destination before it was transmitted.

Perhaps the best way to explain why this makes the field concept necessary would be to give an example. Suppose that I suddenly cause a force to be exerted upon an electron, and that this force causes it to be accelerated. The change in the electron's motion will then cause changes in the electric and magnetic forces that the electron exerts on other charged particles in its vicinity. But if no causal influence can travel at a velocity greater than that of light, this change cannot take place at once. The most reasonable way to describe the interaction between the electron and the other particles is to say that accelerating the electron produces changes in the fields around it, and that these fields carry energy through space and transmit this energy to the other particles.

Gravity provides another example. If gravitational forces can propagate only at velocities no greater than that of light, then one is practically compelled to assume the existence of a field that transmits gravitational energy. It appears that Lagrange's concept of gravitational potential is something more than a mathematical artifice.

It was quantum mechanics that provided the clue to the nature of fields. During the late 1920s, a number of physicists, including Heisenberg, Pauli, the German physicist Pascual Jordan, and the Hungarian physicist Eugene Wigner, began theoretical research on *quantum field theory*. Before long, they had succeeded in showing that all material particles could be understood as the quanta of various fields. According to quantum field theory, there was a field for each kind of particle. There was an electron field, a proton field, and so on.

The idea is not as strange as it seems. Before this group of physicists began their work, it was already known that at least one kind of particle could be described as a manifestation of a field. It was known that light was made up of oscillating electric and magnetic fields, and it was also known that light could be described as a stream of photons. This suggested that photons and electromagnetic fields were different manifestations of the same fundamental phenomenon. The photon could be thought of as a kind of materialization of the electromagnetic force.

But if electrons, protons, and neutrons were manifestations of fields, why were these fields never observed directly the way electromagnetic waves were? The Pauli exclusion principle provided an answer to this question. Since electrons (and also neutrons and protons, which are subject to the exclusion principle too) were prevented from "bunching together," macroscopic electron fields were not seen. Such fields would be observed only if large numbers of electrons with the same energy could act in unison. According to the exclusion principle, this was impossible. Photons, which were not subject to the exclusion principle, behaved differently. Photons possessing the same quantity of energy could propagate through space in a correlated manner.

The discovery that particles could be described as manifestations of fields would not have led to any new discoveries if physicists had been unable to discover theories which described the ways in which different fields interacted with one another. After all, if we say, for example, that an electron can be described as a field, we have really not discovered much that is new. At best, such a theory only provides us with another way of looking at the electron.

However, during the late 1920s and early 1930s, a group of physicists, which included Heisenberg, Pauli, Jordan, Wigner, Dirac, the American physicist J. Robert Oppenheimer, and the German physicist Victor Weisskopf, developed a theory known as *quantum electrodynamics*, which described interactions between charged particles, such as the electron, and electromagnetic fields.

Like Dirac's theory of the electron, quantum electrodynamics—which is often abbreviated by the letters QED—combined quantum mechanics and special relativity. But it went somewhat farther than its predecessor. The behavior of the electron was not only described relativistically, but interactions with the electromagnetic fields were no longer left out.

These interactions are important even in the case of an electron that is spatially separated from other charged particles, for the electron interacts with its own field also. It is a charged particle. Consequently it is surrounded by an electric field. A magnetic field is also created if it happens to be moving. A charged particle and the fields that the charge creates cannot be separated from one another.

If the electron is part of an atom, there are additional interactions to be taken into consideration. An electron is attracted to the positively charged atomic nucleus. In order to move farther away from the nucleus, it must absorb a quantity of energy. This energy is given up in the form of a photon when the electron jumps back down into the energy level that it occupied previously. The emission and absorption of photons of light are a consequence of the interaction between the electron and the electromagnetic fields within the atom.

Quantum electrodynamics did much more than give physicists insights into the emission and absorption of light, however. In fact, it was QED which finally answered the centuries-old question about the nature of the force between particles. In order to see how the theory accomplished this, however, it is first necessary to know something about one of the fundamental principles of quantum mechanics, *Heisenberg's uncertainty principle*.

According to this principle, which Heisenberg formulated in 1927, the position and momentum of a quantum particle can never be simultaneously determined. There is always some uncertainty in our knowledge of one or the other, or of both. More precisely, if one multiplies the uncertainty in a particle's position by the uncertainty in its momentum, the result will be approximately equal to a small quantity called *Planck's constant*, which has a value of 6.6×10^{-27} erg-seconds.

It is really not necessary to know what an erg-second is in order to interpret the principle (although it might be mentioned that the erg is a small unit of energy equal to about two hundred-billionths of a food calorie). The only really important point is that the product of the uncertainties is a small but finite number. This implies that if the uncertainty in the particle's position is somehow made smaller, the uncertainty in the momentum becomes correspondingly larger. Or, if momentum is somehow determined with great accuracy, our knowledge about the particle's position is correspondingly less. The better we know one quantity, the less we can know about the other.

Heisenberg illustrated the principle with a "thought experiment" which can be described as follows: Suppose that one wanted to determine an electron's position very accurately. How would one go about doing this? Well, one way would be to illuminate the electron with radiation of very short wavelength (such as ultraviolet radiation or X rays), and to use the reflected radiation to take a picture. This experiment cannot be carried out in practice, but we need not bother ourselves about that. The purpose of inventing a thought experiment is to gain insight into some aspect of physical reality. It makes little difference if the "experiment" cannot actually be performed.

Quantum theory tells us that short-wavelength radiation is made up of photons of high energy. Thus if we use short-wavelength radiation to illuminate the electron, the electron will be subjected to strong impacts. It will be kicked away at an unknown velocity in an unknown direction. Thus, if we succeed in photographing the electron in a precise position at a given moment in time, any information we might have about the particle's momentum will be destroyed.

We might decide to try illuminating the electron with radiation made up of low-energy photons in order to leave its momentum undisturbed. But if we do this, we cannot accurately determine the electron's position. Low-energy photons are associated with light of long wavelengths. If the wavelength of the illumination used to photograph an electron is much larger than the electron itself, however, the best that one can do is to obtain a very fuzzy picture. It seems that we must conclude that if the momentum (which was presumably determined previously in one way or another) is to be left undisturbed, we cannot obtain a clear picture of the electron's position in space.

In a way, Heisenberg's thought experiment is a bit misleading. It seems to imply that the uncertainty principle does nothing more than place limits on the possible accuracy of our observations. However, in the view of the majority of physicists, it has a much deeper significance. According to the most commonly accepted interpretation, Heisenberg's uncertainty principle implies that a quantum particle does not possess a definite position and a definite momentum at the same time. It can have one or the other, but not both. In other words, a subatomic particle does not behave like a macroscopic object, for which position and momentum can be simultaneously defined.

Of course one could apply the uncertainty principle to macroscopic objects if one wanted to. However, the large size of these objects would make the effects of the uncertainty principle insignificant. For example, suppose the uncertainty in the velocity of an electron was plus or minus one centimeter per second (since momentum is just the product of mass and velocity, the uncertainty in velocity and uncertainty in momentum are related). It is easy to calculate that the uncertainty in the electron's position would be about 7 centimeters, or approximately 3 inches. But if a bowling ball had a velocity that was uncertain to plus or minus one centimeter per second, the uncertainty in its position would be about 10^{-31} centimeters. This quantity is many orders of magnitude smaller than the diameter of an atomic nucleus (about 10^{-12} or 10^{-13} centimeters). Consequently it could not possibly be measured. We can safely conclude that the uncertainty principle has little relevance to the behavior of bowling balls. Even so small an object as a cell in a human body or a bacterium is large enough that quantum effects can be safely ignored.

The significance of the uncertainty principle for quantum field theory arises from the fact that the principle can be applied not only to position and momentum, but also to every pair of what are known as *conjugate variables*. Conjugate variables are quantities that are related to one another in a certain mathematical way.

Energy and time are one such pair of variables. Consequently, the uncertainty principle implies that the energy of a particle cannot be precisely defined over very short time periods. During a very short time span, a particle can have any energy at all. In fact, if the time span is short enough, the energy uncertainty may be great enough to create new particles out of nothing.

Such particles are called *virtual particles*. They spring into existence for tiny fractions of a second, deriving the energy required for their existence from Heisenberg uncertainties. Naturally they cannot exist for long, for this energy "debt" soon has to be paid back. When it is, the virtual particles disappear into nothingness again.

Particles which have a more or less permanent kind of existence, such as electrons in atoms, are called *real particles*. However, this terminology is a bit misleading. Virtual particles, during the short time that they exist, are just as real as any other particle. A virtual electron, for example, does not possess any properties that would distinguish it from one of the real variety.

According to quantum field theory, virtual particles are constantly coming into existence everywhere, even in empty space. Even a perfect vacuum is not devoid of matter; on the contrary, it is an arena in which particles are constantly being created and destroyed. According to quantum field theory, there is no such thing as "nothing."

Virtual particles can also be emitted and absorbed by real particles. It is this latter phenomenon that is responsible for that quantity we call "force." Thus virtual particles provide us with an explanation for the existence of forces and fields.

Quantum electrodynamics is a field theory which applies this concept to the behavior of the electron (and, by implication, to the behavior of other charged particles). QED is based on the assumption that electrons can spontaneously emit virtual photons, and that these photons can be absorbed by other charged particles. Since the photon is the quantum particle associated with the electromagnetic field, the exchange of photons should give rise to electromagnetic forces.

The nature of these forces can be illustrated by means of an analogy. Suppose that two skaters are standing on a frozen lake and that they are throwing a heavy ball back and forth. As either one throws or catches the ball, the ball's momentum will push him away from the other skater.* In this case, they experience something analogous to the repulsive force between a pair of negatively or positively charged particles, such as two electrons (like charges repel one another).

Now suppose that the skaters turn their backs to one another and throw a boomerang back and forth. Since the boomerang changes direction before it is caught, throwing or catching it will push the skaters closer together. This is analogous to the attractive force which exists between a positively and a negatively charged particle, such as an electron and a proton (opposite charges attract).

As with all analogies, this one should not be taken too literally. Although photon exchange can be used to explain electrical attraction and repulsion, it would be a mistake to assume that the virtual photons really follow paths like those of the ball and the boomerang. In fact, Heisenberg's uncertainty principle prevents

---

* It is easy to see why there will be a push when a skater catches it. That there must also be a push when he throws it is a consequence of Newton's law of action and reaction. The thrower of a ball will experience a bit of "recoil."

us from saying anything definite about the path that a photon may travel from one particle to another. In order to determine the photon's trajectory, we would have to know its position and momentum at every instant in time. When we deal with quantum particles—virtual or real—it is never possible to speak meaning-fully of the specific paths that they follow through space and time.

A skeptic might ask whether virtual photons can really be said to exist. After all, he might point out, virtual particles cannot be observed directly. For that matter, we cannot specify the paths they follow in space, or even say precisely how long they exist. According to quantum mechanics, trajectory and time of exis-tence are described by probabilities. For that matter, our skeptic might add, how can one have any faith in a theory which depends upon a belief in the existence of such elusive objects?

One might answer such objections by replying that although virtual particles cannot be seen, their existence can be inferred, just as the existence of atoms is inferred from observations of the Brownian movement and from other experimental data. Virtual photons have real observable effects, which can be detected in experiments. In fact, QED is one of the most accurately con-firmed theories in physics. Experiments have been performed which verify its predictions to an accuracy of better than one part in a billion. For example, quantum electrodynamics tells us that the magnetic field of an electron should be exactly 0.1159652 percent stronger than the value that is given by the Dirac theory. This is exactly what is observed.

And what does the magnetic field of a spinning electron have to do with exchanged photons? Well, if the theory is correct, an electron also exchanges photons with itself. There is no require-ment that an emitted photon must be absorbed by a different particle; it can be reabsorbed by the same one that emitted it.

When one speaks of the emission and absorption of photons, the picture that one is painting is a simplified one. There is noth-ing that requires a virtual particle to retain its identity during the short period that it exists. As long as certain basic laws of physics are followed, the virtual particles may undergo endless series of transformations. For example, an electron may emit a photon. The photon may decompose, in turn, into an electron-positron pair. The members of this virtual pair can emit additional photons. Alternatively, the electron and positron may annihilate one another, transforming themselves into a virtual photon again.

An electron (or other charged particle) thus surrounds itself with a cloud of virtual particles of many different kinds, including photons, electrons and positrons. The cloud also contains a few heavier particles, such as virtual protons and antiprotons. However, since more energy is required to create pairs of heavy particles, they are present in correspondingly smaller quantities.

The presence of charged virtual particles around an electron alters certain of the electron's observable characteristics, including its magnetic field. Changes in the magnetic field affect the interaction between the electron and an atomic nucleus. This changes atomic energy levels slightly. These energy shifts alter the wavelengths of the light or other radiation that an atom may emit. Admittedly, going from the existence of virtual particles to changes in the wavelengths of emitted light involves a long chain of reasoning. However, each step is straightforward, and there are no other explanations for the shifts that are observed.

There is just one problem associated with this picture. When one computes the energy of the interaction between an electron and the cloud of virtual particles surrounding it, one obtains an infinite result. For that matter, QED also predicts that the energy of "empty" space should be infinite. The energy of the interactions between the virtual particles that pop into existence in the absence of matter can be computed in the same way that the interaction between an electron and its own field is calculated. When this computation is carried out, one obtains the same kind of undesirable result.

When a calculation gives an infinite result, this is generally an indication that there is something wrong with the theory that one is using. After all, infinite quantities are not observed in nature. When one does encounter infinite mathematical quantities, it is impossible to assign any meaning to them.

Indeed, throughout the 1930s, the accepted belief among physicists was that quantum electrodynamics suffered from serious internal inconsistencies, and that drastic modifications would have to be made in the theory before it would make any sense.

Then, in 1936, Weisskopf suggested a way of getting around

these difficulties by use of a mathematical trick. During the 1940s, Weisskopf's method was developed further by the Japanese physicist Shin'ichiro Tomonaga and by the Americans Julian Schwinger and Richard Feynman. The technique, which was called *renormalization*, depended upon the apparently questionable mathematical procedure of subtracting certain infinite quantities from other infinities. Finite results were obtained, and it appeared to be possible to do meaningful calculations.

The reason that the renormalization procedure seemed to be questionable was that the quantity infinity minus infinity can be equal to anything. As every mathematician knows, one cannot juggle infinite quantities in this matter. Yet, somehow, the renormalization procedure worked. Calculations that were done within the framework of renormalized quantum electrodynamics gave numerical results of unprecedented accuracy.

Some physicists remained unimpressed even when QED yielded results that were confirmed to accuracies that had rarely been achieved before in physics. For example, Dirac has said of the renormalization procedure, "This is just not sensible mathematics. Sensible mathematics involves neglecting a quantity when it turns out to be small—not neglecting it because it is infinitely great and you do not want it!" Dirac feels that the problem with infinities is an indication that drastic changes will have to be made in quantum electrodynamics before it can be considered to be a satisfactory theory.

Other physicists are more sanguine. Many of them find that they have no qualms about renormalization. In their view, the fact that the method works is an indication of its validity. These physicists have not hesitated to extend the method of renormalization and make use of it in other types of quantum field theory.

There may very well be a sense in which renormalization is mathematically acceptable. It is possible that mathematicians may eventually find some way to justify the procedure. This sort of thing has happened frequently in physics. There have been numerous occasions on which theories have made use of mathematical procedures that seemed to be illogical. In most of these cases, physicists and mathematicians have found ways to "clean up" the mathematics and cause it to adhere to the requirements of consistency and logic. In fact, calculus itself, the branch of mathematics upon which all of physics is based, was thought to be logically inconsistent for more than a century after it was discovered. Yet ways were eventually found to put calculus upon a formal logical foundation.

Although the renormalization procedure might eventually turn out to have more validity than some of the skeptics suggest, there is a sense in which renormalized QED fails to give us an understandable picture of the nature of subatomic reality. Its numerical predictions may be accurate. However, it seems to imply that the electron has properties which are somewhat paradoxical.

If we take the implications of quantum electrodynamics literally, we are forced to conclude that an electron is an object that possesses infinite mass, infinite charge, and infinite energy. The reason that these infinite quantities are never observed is presumably that the electron is screened by the cloud of virtual particles that surround it. One never sees a "bare" electron; there are always numerous virtual particles that get in the way.

The renormalization procedure does account for the screening. For example, consider the fact that a "bare" electron supposedly possesses infinite energy. The interaction between the electron and its own field (i.e., its cloud of virtual particles) is infinite also. These two infinities cancel one another out, so that when one looks at the combination electron-plus-field, one observes only a finite energy. Renormalization causes the infinite mass and infinite charge to be canceled out in the same way.

What, then, is an electron? We must admit that we do not know. Perhaps QED is telling us that there is no sensible way of thinking of an electron apart from the field that surrounds it. Or perhaps the drastic changes in the theory that Dirac has called for will eventually come about. In that case, some new and surprising model of the electron may be developed.

At the moment, there is only one thing that seems to be clear. Whatever the electron is, it is not the tiny, solid particle that J. J. Thomson envisioned. Though one speaks of "electrons" today, using the same word that Thomson used in the 1890s, the meaning of the term has changed so much that one cannot claim to be speaking about the same thing. Thomson thought of the electron as a tiny piece of charged matter. To the contemporary physicist, on the other hand, it is a mysterious entity that refuses to reveal its true nature, and remains hidden behind the cloud of its own self-field.

If quantum electrodynamics does not satisfy our desire to know what an electron really "is," and if it sometimes causes us to wonder if that question even has meaning, QED and other quantum field theories have enabled us to obtain a clearer picture of other aspects of physical reality. In particular, the development and elaboration of quantum field theories have led to progress toward an understanding of the nature of force, and toward a solution of the old problem of action at a distance.

If quantum field theory is correct in its broad outlines, as it seems to be, then it is not hard to understand why instantaneous action at a distance cannot exist. Forces propagate no faster than the speed of light because force is a phenomenon that depends upon the exchange of particles. Electromagnetic forces can be explained by photon exchange. Gravity is presumably the result of the exchange of certain as-yet undiscovered particles called *gravitons*. Finally, the two different kinds of nuclear force that physicists have discovered (these will be discussed in the next chapter) are the result of the exchange of yet other kinds of virtual particles.

Photons can travel at the speed of light (it would be a bit paradoxical if they could not; after all, they *are* light) because they have zero rest mass. If a particle has no rest mass, its motion will not be hindered by the relativistic mass increase. Gravitons are thought to be massless also. If they are, they too travel at light velocity. In general, the virtual particles associated with nuclear forces are not massless. But they travel such short distances that it would not be meaningful to speak of their "velocities" anyway; we would encounter the usual difficulties associated with the uncertainty principle.

Quantum field theory not only explains the nature of forces and fields, but it also gives us insights into the nature of matter. Perhaps one encounters difficulties when trying to understand what an electron is "really" like. Nevertheless, there are many kinds of real insight that can be gained. Some of these will be discussed in the following chapter.

3

The Nature of Matter

Today, physicists often say that the universe is constructed of quantum fields. Subatomic particles are nothing more than manifestations of fields; nothing else exists. Their reason for saying this is that scientists prefer to explain complex phenomena in terms of ideas that are relatively more simple. It is mathematically simpler to describe matter as a set of fields than it is to deal individually with numerous different kinds of particles.

However, with the exception of the electromagnetic and gravitational fields, it is particles that are seen in the laboratory. No one has ever observed an electron field, for example. As a result, attempts to understand the nature of matter have centered upon observations of the behavior of electrons, protons, neutrons, and the numerous other kinds of particles that can be created in modern high-energy accelerators. J. J. Thomson studied particles during the years in which he was trying to understand the nature of the electron, and observations of the behavior of particles are still yielding significant results today.

The field of particle physics can be said to have begun with J. J. Thomson's discovery of the electron in 1897. In the years that followed, one discovery was made after another. By around 1920, it appeared that the fundamental constituents of matter were known. Rutherford had discovered the atomic nucleus in 1911, and in 1919, he had announced his discovery that when the nuclei of light atoms were bombarded with alpha particles, those nuclei broke apart. Hydrogen nuclei—later to be named "protons"— were found among the disintegration products. This implied that the nuclei of all elements must be made of protons. If this was true, then it followed that electrons and protons must be the fundamental constituents of matter. These two particles, together with photons, could be used to explain all the phenomena that physicists observed. Or so it was thought.

The picture of the atom that evolved during the second decade of the twentieth century can be best described by giving a few examples. The lightest element was hydrogen. A hydrogen atom was made up of an electron and a proton. The two particles had equal and opposite electrical charges, so the atom itself was electrically neutral.

The next-simplest atom was helium. The helium nucleus was about 4 times as heavy as that of hydrogen. However, it had a positive charge of only two units, not four. Hence it seemed reasonable to assume that the nucleus must contain four protons and two electrons. The negatively charged electrons would cancel out two of the positive charges, giving a net electrical charge of $+2$. Since two electrons circled the helium nucleus, the net result would again be an atom with zero net charge.

It appeared that all of the other elements could be built up in the same manner. An oxygen nucleus, for example, had a charge of $+8$ and a weight of 16. Obviously it contained sixteen protons and eight electrons. Eight additional electrons revolved around this nucleus. Apparently, then, an oxygen atom contained sixteen protons and sixteen electrons; eight of the latter were in the nucleus and eight occupied various orbits.

At the time, there seemed to be experimental evidence to indicate that this model was correct. It was observed that many radioactive elements emitted negatively charged objects called *beta particles*. It was clear that these particles were nothing but high-velocity electrons; it had already been shown by 1900 that beta particles and Thomson's cathode ray particles had identical properties. Furthermore, there was evidence that the beta particles were emitted from the nucleus. While he was still an assistant of Rutherford's, Bohr had found theoretical arguments which indicated that electrons emitted in beta decay had a nuclear origin.

One might think that it would have been a simple matter to test the theory that nuclei were made up of protons and electrons. In the case of helium, for example, couldn't one simply add the weights of the nucleus' constituents—four protons and two electrons—and then check to see if their combined mass was equal to that of the nucleus?

Such measurements could be performed during the 1920s. However, the results were not easy to interpret. According to Einstein's equation $E = mc^2$, the energy that bound the constituents of a nucleus together contributed to the nuclear mass. Since, at the time, nothing was known about the forces that bound these constituent particles together, it was impossible to calculate the binding energy, or its equivalent mass. No one could tell whether nuclei had the theoretically correct masses or not.

On the other hand, there existed no evidence against the idea that matter was made up of electrons and protons. Physicists continued to accept this theory until quantum mechanics was developed in the mid-1920s. And then they suddenly discovered that something was terribly wrong. According to Heisenberg's uncertainty principle, electrons could not possibly remain confined within an atomic nucleus.

The argument went something like this: Nuclei were very small; they had dimensions of about 10^{-12} centimeters. Therefore, if an electron were confined within a nucleus, its position would be known very exactly. This implied that there must be a large uncertainty in the electron's momentum, and hence in its velocity. This implied, in turn, that the electron spent a considerable part of its time moving very rapidly; it would often have a speed that was more than 99.9 percent of the velocity of light. But an electron that was moving that fast could not remain confined within a nucleus; it would break away from the nuclear protons in a small fraction of a second. According to quantum mechanics, all elements had to be constantly emitting beta particles if the accepted theory of nuclear structure was correct.

If the uncertainty principle did no more than place limits on the possible accuracy of measurements, this argument might not seem very compelling. But, as we have seen previously, the principle accomplishes much more than that; it can be used to deduce the existence of virtual particles, for example. Hence it should come as no surprise that it can be used to demonstrate the impossibility of the confinement of electrons within a nucleus. Incidentally, this argument does not apply to the proton. Protons can remain confined because they are so much heavier than electrons. Because they have more mass, they have velocities that are much smaller for any given value of their momentum. Protons and electrons can be likened to such objects as boulders and pebbles. A pebble would have to travel rapidly indeed if it were to have the same momentum as a slowly moving boulder.

This argument put physicists in a quandary. On one hand, there was evidence which indicated that electrons could be emitted from nuclei. On the other hand, the uncertainty principle— which was one of the fundamental tenets of quantum mechanics—said that electrons could not be nuclear constituents.

The problem was cleared up in 1932, when the English physicist James Chadwick discovered a third fundamental particle, called the *neutron*. The neutron had a mass which was just a bit more than that of the proton. The most significant difference between the two seemed to be that the neutron did not have any electrical charge.

Heisenberg then suggested that nuclei were made up not of electrons and protons, but of protons and neutrons. For example, the helium nucleus contained two protons and two neutrons. The oxygen nucleus, which had previously been thought to contain sixteen protons and eight electrons, could be a combination of eight protons and eight neutrons. The assumption that matter was made up of three elementary particles was, to be sure, a little more complicated than the hypothesis that there were only two. However, everything seemed to work out correctly.

Or, at least, almost everything added up. There were still one or two problems. For example, it had been known for some time that the beta particles that were emitted from nuclei did not always possess the same energy. Sometimes they would have a bit more than the average, sometimes a bit less. Even when one observed beta emission from nuclei of the same type, the energy was not constant. It appeared that some of the beta particles' energy was disappearing.

Another problem had to do with the spin of the neutron. Like the electron and the proton, the neutron had a half-quantum of spin.* This was in itself no problem. In fact, if neutrons and protons resembled one another, one might expect that their spin would be the same. However, problems arose when one considered the fact that a free neutron—one that was not bound in a nucleus—would decay into an electron and a proton in about twelve minutes.

The decay of the neutron was not intrinsically puzzling either; it seemed to be an analogue of the beta decay processes that took place within nuclei. The problem was that the spins of the particles did not add up. When two spin-½ particles are combined, the spins must either be added to one another or subtracted, giving a result of either 1 or 0. In other words, only spinless particles or particles possessing one quantum of spin should be able to decay into two particles of spin ½.

A possible solution to the dilemma was found in 1930, two years before the neutron was discovered. In that year, Wolfgang Pauli pointed out that the problem of the missing energy in beta decay processes could be solved if one assumed that two particles, rather than one, were being emitted. If some of the energy was carried away by a new particle—which the Italian physicist Enrico Fermi later named the *neutrino*—this would explain why the energy of the electrons emitted by radioactive nuclei was not always the same.

At first, Pauli had some doubts about the hypothesis, and he did not fully commit himself to it. However, when the neutron was discovered, he saw that his theory would also clear up the problems concerning neutron spin. A spin-½ neutron should not be able to decay into two particles that had a half-quantum of spin

* A quantum of spin is defined to be the quantity $h/2\pi$, where h is Planck's constant and π is the Greek letter pi, which denotes the ratio of the circumference of a circle to its diameter.

each. But if there were three spin-½ particles, there was no problem. If two of them spun in opposite directions, their spins would cancel out (½ − ½ = 0). Thus if the neutrino was also a spin-½ particle, one could have a total spin of ½ both before and after decay.

Pauli knew of no way in which the spin of the neutrino could be experimentally measured. For that matter, it was not obvious that the neutrino could even be detected. Since it had no charge, it would not be deflected in an electric or magnetic field, and since its mass had to be much less than that of an electron, it would be extremely difficult to observe its interactions with other particles.

After Chadwick had discovered the neutron, Pauli realized that there might be more to his neutrino hypothesis than he had originally thought. So he proposed the idea again, this time with somewhat more confidence, to a gathering of physicists in 1933. Later that year, Fermi worked out a theory of beta decay in mathematical detail, showing that the neutrino hypothesis was a reasonable one indeed.

It should be pointed out that none of this implied that the neutron was a composite particle. On the contrary, it was as elementary as the proton and the electron. When neutrons decay into electrons, protons, and neutrinos, they are not breaking apart into their constituent parts. The neutron ceases to exist, and the other three particles appear in its place. That something like this should happen should not be considered to be very surprising. Such transformations are common in the subatomic world. As we have seen previously, an electron can emit a photon, and a photon can transform itself into an electron and a positron. The creation of three particles from one is only a bit more complicated.

Since the time that Pauli proposed his beta decay hypothesis, there has been a slight change in terminology. Today we say that the neutron decays into a proton, an electron, and an *antineutrino*. The antineutrino is the neutrino's antiparticle. The description of the process is simpler if one assumes that a particle (the electron) and an antiparticle (the antineutrino) are created together. However, the term "neutrino" can be used generically to describe both the neutrino and its antiparticle.

Of all the particles known to physics, the neutrino is the most elusive. Neutrinos have no charge and no mass,* and they travel at the speed of light. According to the special theory of relativity, massless particles must travel at light velocity. If they did not, they would have zero energy, and they could not even be said to exist. Speaking rather loosely, one can say that the relativistic mass increase transforms a particle with zero rest mass into something that possesses momentum, even though its mass can still not be defined.

Neutrinos may be the most common particles in the universe. There are approximately a billion of them for every proton or electron. Yet they interact with matter very rarely. It has been estimated that a neutrino could pass through thousands of light-years of solid lead before it was stopped. Nevertheless, neutrinos can be detected. In 1956, the American physicists Clyde L. Cowan, Jr., and Frederick Reines set up an experiment next to a nuclear reactor that was producing an estimated 10^{18} (a billion billion) neutrinos every second. Although the chance that any given neutrino would interact with the two physicists' detector was extremely small, the particles were present in such great numbers that some neutrino interactions could be observed. Today, neutrino interactions are frequently observed in experiments performed on high-energy particle accelerators.

There is much more that can be said about the behavior of neutrinos, and I will make reference to the subject from time to time. At the moment, however, it might be best not to stray too far from the subject of physicists' changing conceptions of the nature of matter.

In 1920, scientists knew of two fundamental constituents of matter, the electron and the proton. By 1935, there were five; the proton, the electron, the positron, the neutron, and the neutrino. Of course, a sixth particle, the photon, was also known. But the photon was not really a constituent of matter; it was a particle that was associated with the electromagnetic interaction.

Matters were not as simple as physicists had initially believed them to be. On the other hand, things had not yet grown unduly

---

* There exists some experimental evidence which indicates that the neutrino might have a small mass—a tiny fraction of that of an electron. However, the experiments which indicate this are still quite controversial.

complicated. Atoms, after all, appeared to be made up of only three of the particles. Neutrinos were presumably created only in radioactive decay processes, and the positron had been seen only in cosmic ray showers.

Then, in 1935, the Japanese physicist Hideki Yukawa pointed out that at least one more particle should exist. If quantum field theory provided an accurate description of nature, then there must exist an as-yet undiscovered particle associated with the nuclear force. This new particle, which Yukawa called the *meson*, played a role analogous to that of the photon. Just as virtual photons were responsible for the electrical attraction that bound electrons and nuclei together in atoms, the exchange of virtual mesons could cause protons and neutrons to be bound together in the atomic nucleus.

Yukawa calculated that the meson had a mass that was roughly 200 times as great as that of an electron, or about one-ninth that of a neutron or proton. The reason that the meson, unlike the photon, had to have a mass was that the nuclear force was observed to have a very short range; it acted only over distances comparable to the size of the nucleus. At greater distances, it rapidly fell off to zero.

It is not very difficult to see why short-range forces must be associated with massive particles. This fact is a simple consequence of the uncertainty principle. Recall that the product of the energy of a virtual particle and the time that the uncertainty principle allows it to exist is approximately equal to Planck's constant. This implies that the more massive (more massive because mass and energy are equivalent) virtual particles can exist only for relatively short periods of time. But if a particle does not exist for very long, it will not be able to travel very far before it is annihilated. Therefore massive particles create short-range forces.

There is also an analogy that can be used. The explanatory power of the analogy is not very great. However, familiarity with it can help one remember the relationship between the range of a force and the virtual particles' mass. Suppose that two skaters are throwing a very heavy ball back and forth. Obviously, they will not be able to throw it very far. On the other hand, if the ball is light, the forces between them can act over much larger distances. And if the ball has no mass at all, there will be no limit to the distance

that the skaters can throw it. Naturally we have to assume that the
ball is not impeded by air resistance. But this is reasonable
enough when one is trying to make an analogy with the behavior
of subatomic particles.

In the same year that Yukawa put forth his hypothesis of
nuclear force, Carl Anderson discovered a particle that seemed to
be the predicted meson. The particle, now called the *mu* meson*,
was discovered in cosmic ray showers. It had the same electric
charge as the electron, and it appeared to have approximately the
mass that Yukawa had predicted.

Today, this particle is called the *muon*. It is not really a meson
at all. Shortly after Anderson had made his discovery, physicists
realized that the muon was not the particle that Yukawa had
predicted. On the contrary, it was something which, except for its
mass, bore a very great resemblance to the electron.

Yukawa's particle, called the *pi† meson*, or *pion*, was finally
discovered in 1947. There are three varieties of this particle: the
pion can have either a positive or a negative electric charge, or it
can be neutral. Unlike the muon, which has a spin of ½, the spin
of the pion is zero. The spin of a pion can be determined without
much difficulty because it is not always only a virtual particle. Real
pions can be created if enough energy is available.

It is the spin of the pion that tells us that it is associated with a
force, and the spin of the muon indicates that it cannot be. The
particles that are observed in nature can be divided into two categories, *fermions* and *bosons*. They are named after Fermi and after
the Indian physicist Satyendra Bose, respectively. Bosons are particles which possess zero spin, or one quantum of spin, or two
quanta. They are associated with forces. Fermions have half-integral spin (½ or ³⁄₂ or ⁵⁄₂, and so on). They are particles of matter.

Naturally, the definition of a quantum of spin is somewhat
arbitrary. If it were defined differently, the distinction between
fermions and bosons would have to be made in a different way.
For example, if the unit of spin were taken to be half as big, we
would say that bosons had a spin that was odd and fermions a

* Mu is one of the letters of the Greek alphabet.
† Another Greek letter, the one that is used in another context to give the
circumference of a circle.

spin that was even. The half-integral spin of fermions, in other words, is the result of a certain convention. It does not imply that there is any such thing as "half of a spin."

Another distinction that can be made between fermions and bosons is that the former obey the exclusion principle, while the latter do not. This fact is related to certain mathematical properties of the probability waves that describe these two different kinds of particles.

The fact that a muon has a spin of ½ implies that it cannot be associated with a force. In fact, the muon can be described as a kind of "fat electron." It has electronlike properties, although it is about 200 times as heavy.

The forces that bind protons and neutrons together in nuclei are created by pion exchange. This phenomenon is a bit more complicated than the photon exchange that creates electromagnetic forces. Again, it might be best to attempt to explain it by means of examples. For simplicity, I will consider the case of a nucleus that contains only one proton and one neutron. The exchange process in more complicated cases is analogous.

Either the proton or the neutron may emit an uncharged pion, which is then absorbed by the other particle. This will give rise to an attractive force. However, it is not the only kind of exchange that can take place. For example, the proton can emit a positively charged pion, transforming itself into a neutron in the process. The positive pion can then travel to the neutron, changing it into a proton. This exchange not only produces forces that bind proton and neutron together, it also causes them to exchange identities. Similarly, the neutron can emit a negative pion. When it does, it acquires a positive charge and becomes a proton. Then, when the proton absorbs the pion, it becomes a neutron. Again, the proton and the neutron trade roles.

The fact that such metamorphoses can take place has induced physicists to invent the term *nucleon* to describe both the proton and the neutron. Proton and neutron can be considered as two different states of the same particle.

With the discovery of the pion in 1947, it appeared that physicists had at last begun to attain an understanding of the nature of matter, and of the forces which existed between elementary particles. And then, almost at once, new problems began to appear.

Studies of cosmic ray showers turned up new kinds of particles. Beginning in 1948, experiments with cyclotrons produced additional kinds of particles in the laboratory. It was discovered that there were numerous different varieties of mesons, and a large number of different kinds of *baryons*—heavy particles that bore a resemblance to the proton and neutron. No one could explain why so many different kinds of particles should exist.

For that matter, physicists could not even explain the existence of the muon, which had been known since 1932. It was clear enough that the muon was a kind of heavy electron. However, it was not a constituent of ordinary matter, and it seemed to play no significant role in the scheme of things. The muon decayed into an electron in approximately two millionths of a second. Physicists realized that if all the muons in the universe were suddenly to disappear, no one but experimental physicists would ever know the difference.

Another problem had to do with the fact that it seemed impossible to formulate a quantum field theory which described the nuclear force. Analogies with quantum electrodynamics had suggested to Yukawa that the pion ought to exist. But no one knew how to work out a theory of the pion in detail, and physicists were unable to determine exactly how the nuclear force varied with distance. The difficulties were so great that some of them even began to suggest that there might be circumstances under which the ideas of quantum field theory would have to be abandoned.

By the end of the 1950s, so many different particles had been discovered that it was obvious that the term "elementary particle" was a misnomer. There were apparently hundreds of different kinds of mesons, and hundreds of different kinds of baryons as well. Furthermore, there was every reason to believe that hundreds, perhaps thousands, of additional particles would be discovered in the future. It was even possible that these "elementary" constituents of matter could turn out to be infinite in number.

As we have seen, the fundamental concepts of physics have had a tendency to undergo transformations as new theories have been developed. The "electron" of quantum field theory bears little resemblance to the "electron" that J. J. Thomson envisioned. The concept of energy changed when Einstein showed that matter and energy were equivalent. Originally, the concept of force was based on the naive idea of a push or a pull. This was modified when Newton introduced the idea of action at a distance. It was modified even more dramatically when the physicists who developed quantum electrodynamics showed that force could be understood as an exchange of virtual particles.

However, in every case, the same basic ideas were retained. Scientists spoke of "force" in Newton's time, and they still do so today (although they sometimes prefer to use the equivalent term "interaction"). One can still speak of "matter" and "energy" and perform mathematical calculations that make use of these concepts. Accepted ideas about the nature of the electron may bear little resemblance to those which were in vogue in 1897, but when physicists speak of "electrons," no one suggests that the term no longer has meaning.

The situation was quite different in the field of particle physics. Scientists had set out to discover what the fundamental constituents of matter were. By the end of the 1950s, they were beginning to wonder if there really were any. They had begun with an idea that was relatively simple—that matter was made of a few elementary particles—and had encountered an ever-worsening chaos as they had attempted to work out its consequences.

Physicists were dismayed at the complexity that they had uncovered. Physical science has always been based on the idea that natural phenomena can be explained in terms of simple principles. When a theory becomes too complicated, physicists have a tendency to distrust it. There is really no way to prove that nature is fundamentally as simple as scientists like to think. It could be argued that the demand for simplicity is nothing more than a philosophical prejudice. However, if it is a prejudice, it is one that has proved to be very useful.

The Copernican revolution came about because astronomers such as Copernicus, Kepler, and Galileo felt that the old earth-centered astronomy was just too complicated to be true. If one wanted to assume that the earth was really the center of the solar system, it was necessary to assume that planets moved in epicycles. But this was only the beginning. When the system was worked out in detail, one had to assume that there were epicycles upon epicycles. Even this was not always enough. Astronomers also had to invent techniques that made use of mathematical quantities known as equants and deferents. In effect, this amounted to assuming that certain planets did not move around the earth after all, but around points in space near the earth that were themselves moving. The heliocentric theory eventually won out because it was much simpler, and so transparently logical that it was difficult to doubt that it was true.

Similarly, Newton's inverse-square law of gravitation was accepted because it gave such a simple description of the workings of the universe. According to Newton, gravity was a universal force that explained the motions of celestial objcts and terrestrial bodies alike. There was just one kind of gravity. It was not necessary to use one theory to explain the behavior of falling bodies, and another to describe the motions of the planets.

The special theory of relativity represented a step toward simplicity too. It was based on two principles, each of which was simplicity itself: that the speed of light was constant, and that there were no ways to distinguish between relative and absolute motion. Once these assumptions were made, it became possible to discard all the convoluted theories of the ether that physicists had worked so hard to elaborate.

Physical theories sometimes become mathematically very complex. There are problems in Newtonian physics that are so difficult that they have never been solved. For example, no one has ever found an exact solution to the equations that describe the motions of three gravitating bodies (for example, the sun, the moon and the earth) in their mutual gravitational fields. Indeed, an exact solution may not even exist. The best that scientists have been able to do is to find approximate methods of solving this three-body problem.

However, complex phenomena and complex fundamental principles are two different things. Scientists are constantly seeking ways to deal with the former while they eschew the latter. In fact, it would not be inaccurate to define science as the attempt to discover the fundamental simplicity that is hidden behind the apparent complexity of natural phenomena.

Therefore it is not surprising that the physicists who studied the behavior of "elementary" particles should have sought to find new principles that would allow them to make some sense of the chaos that they had uncovered. Nature couldn't be as complicated as it seemed to be.

The first step was to classify particles according to the forces that they experienced. By the end of the 1950s, it had been known for some years that the interactions between the particles that were observed in nature depended upon four fundamental forces; the gravitational interaction, the electromagnetic interaction, and the *strong* and *weak* nuclear forces.

Attention was concentrated upon the last two. It was obvious that gravitation was much too weak a force to have much significance for the behavior of elementary particles. A simple calculation showed that the strong force between a pair of protons, for example, was approximately 10^{39} times greater than the protons' gravitational attraction. It appeared that gravity was important only when one dealt with bodies of large mass.

The electromagnetic interaction had greater significance for particle physics. But it had long been understood. The observation that electromagnetic forces acted only between particles that had electric charges or magnetic fields was too obvious to be very useful. On the other hand, the fact that some particles were subject only to the weak interaction, while others felt both the strong and weak forces, was something that might have some fundamental significance.

The strong force is the force that binds protons and neutrons together in a nucleus. It is the one that I have been calling the "nuclear" force up to now. The weak force has a somewhat different character. It does not cause particles to stick together or to be pushed apart as the strong and electromagnetic interactions do. The weak force is responsible for certain types of interaction, such as beta decay. It has an intrinsic strength that is about a million times less than that of the strong force, and it acts at correspondingly small distances. Where the strong force has a range of about 10^{-13} centimeters, that of the weak force is 10^{-16} centimeters, about a thousand times less.

The existence of two distinct nuclear forces suggested that elementary particles could be divided into two different families. Particles that felt the strong force were called *hadrons*. The hadron family was made up of baryons and mesons. Particles that felt the weak force, but mot the strong one, were called *leptons*. The lepton family was made up of the electron, the muon, the neutrino, and their antiparticles. Since the classification was devised, a new particle, called the *tauon* (tau is another Greek letter) has been added, and experiments have shown that there are three different kinds of neutrino.

If the muon is a kind of heavy electron, the tauon can be thought of as a superheavy electron. It weighs about 3500 times as much as the electron, or about 170 times as much as the muon. The three kinds of neutrino are the electron neutrino, the muon neutrino, and the tauon neutrino: each of the electronlike particles has its own neutrino partner.

At this point, names for so many different kinds of particles have been introduced that it might be a good idea to recapitulate a bit in order to avoid confusion.

Hadrons are particles which feel the strong force. The family is divided into baryons (heavy particles similar to the neutron and proton) and mesons (for example, the pion). It might be noted that hadrons are subject to the weak force also. For example, it is the weak force which causes neutrons to decay.

Leptons are particles which do not feel the strong force, but which are subject to the weak interaction. The family of leptons has six members: the electron, the muon, the tauon, and the three neutrinos. Antiparticles such as positrons and antineutrinos are also members of the family; however, they are not counted separately.

The photon, which feels neither the strong nor the weak force, is placed in a class by itself, as is the graviton.

One only has to look at the classification scheme to feel that some progress has been made. If there are only six leptons, perhaps that isn't so bad. It could be that all six really are elementary. On the other hand, the hadrons exist in unreasonable numbers. It seems reasonable, therefore, to look for ways of classifying hadrons further. If they can be subdivided into groups, this might be a first step toward bringing order out of chaos.

Such a step was taken in 1961 when the American physicist Murray Gell-Mann and the Israeli intelligence-officer-turned-physicist Yuval Ne'eman independently discovered a scheme, called the *eightfold way*, that allowed baryons and mesons to be grouped into subfamilies. It was given this name because the most commonly observed baryons and mesons were put together in two sets of eight, called the *baryon octet* and the *meson octet*. Of course, the term "eightfold way" is a pun; the original eightfold way was a recipe for enlightenment that was devised by the Buddha in the fifth or sixth century B.C. Gell-Mann and Ne'eman may or may not have felt that they had set particle physics on the path to Nirvana. However, their example has prompted other physicists to coin even worse puns when they needed to invent new terminology. Examples will be encountered later in the book.

The next step, after the eightfold way was devised, was to find out why this particular classification scheme worked. If certain particles exhibited similarities that made it possible to group them together, then there had to be some underlying principle that would explain why the method seemed so natural. A possible way of interpreting the scheme was suggested by Gell-Mann, and independently by the American physicist George Zweig, in 1964. The two physicists pointed out that the eightfold way could be explained if one assumed that baryons and mesons had even smaller constituents.

Zweig suggested that the new subhadronic particles be called *aces*. Gell-Mann named them quarks. It was the latter name that stuck. *Quark*, incidentally, is a German word meaning "curds." Gell-Mann took the term from James Joyce's novel *Finnegan's Wake*. Speaking of the cuckolding of King Mark in the legend of Tristram and Iseult, Joyce says, at one point, "Three quarks for Muster Mark."

Fortunately, the quarks themselves are somewhat simpler than their etymology. In the original version of the theory there were just three of them, called *up, down,* and *strange*. The name "strange" should not be taken literally. "Strangeness" is a whimsical name that is given to a mathematical property possessed by certain particles; it is one of the properties upon which the eightfold-way classification scheme is based.

All particles have antiparticles.* Hence if there were three quarks, there also had to be three antiquarks. These were called

---

* However, there are a few, like the photon, which are their own antiparticles. The photon possesses the odd property of being a particle and an antiparticle at the same time.

antiup, antidown, and *antistrange.* According to the theory, all of the hundreds of known baryons and mesons could be explained as combinations of these six quarks and antiquarks. Baryons were made of three quarks, while a meson was composed of a quark and an antiquark. For example, the proton contained one down and two up quarks, while the constituents of a positively charged pion were an up and an antidown.

One might think that the number of possible combinations would have quickly been exhausted, and that six quarks would not be sufficient to explain all the hadrons that had been observed. But this is not really the case. The three quarks that make up a proton, for example, have a given amount of energy. If additional energy is imparted to them, Einstein's equation $E = mc^2$ requires that the proton gain mass. But if it gains mass, it will not be a proton any longer; it will be one of the protonlike particles that make up part of the family of baryons. And if the extra energy that the quarks have obtained is given up, the particle will become a proton again. Naturally, energy can also be given to the quarks that make up a pion, and new mesons can thereby be created.

The quark theory, in other words, explained why baryons and mesons existed in such great numbers. Many of the "new" particles that physicists had been discovering were nothing more than higher energy states of old ones. They were particles whose constituent quarks had absorbed various quantities of energy.

Quarks were conceived to be fermions with spins of $\frac{1}{2}$. Thus two of them could combine to make a pion of spin zero ($\frac{1}{2} - \frac{1}{2} = 0$), and three could combine to produce a spin-$\frac{1}{2}$ proton or neutron ($\frac{1}{2} + \frac{1}{2} - \frac{1}{2} = \frac{1}{2}$). Finally, according to the theory, quarks had electrical charges that were either one-third or two-thirds of those possessed by particles like the proton or electron. The charges could always be added together so that whole-number charges were obtained. For example, the two up quarks in a proton had charges of $+\frac{2}{3}$ each. This gave $+\frac{4}{3}$. When one subtracted the $-\frac{1}{3}$ charge of the down quark, the result was $+1$. In the neutron, on the other hand, positively and negatively charged quarks combined to give a net electrical charge of zero.

Although everything seemed to work out correctly and although the quark theory seemed to be capable of explaining the properties of all of the particles that had been observed, many physicists regarded the quark theory as nothing more than a convenient mathematical fiction. Their attitude about the reality of the quark was more or less the same as that of Ernst Mach about the reality of the atom. When attempts to find free quarks in nature failed, these physicists' suspicions were confirmed. The concept of the quark, they concluded, was an abstract idea that corresponded to nothing real.

Then, in 1968, an experiment was performed at the Stanford Linear Accelerator Center (SLAC) that cast doubts on this skeptical interpretation. When protons were bombarded with high-energy electrons, pointlike charges were discovered to exist inside the proton. Quarks were apparently very real.

At first it seemed that physicists had achieved a simple explanation of the nature of elementary particles. But then the quarks began to multiply, just as the baryons and mesons had. In 1974, physicists at SLAC and at the Brookhaven National Laboratory discovered a new hadron which could not be explained as a combination of the three known quarks. It became necessary to postulate the existence of a fourth quark, called the *charmed* quark. Subsequently, more hadrons containing this fourth quark were discovered at SLAC and at a German laboratory near Hamburg.

Nor was that the end of the proliferation. A fifth quark, the *bottom* quark, was discovered in 1978. A sixth, the *top* quark, was found in 1984. Naturally it is possible that yet other quarks might exist. There are no known theoretical reasons why the list of quarks should have to come to an end at six. The fact that there are also six known leptons creates an appealing symmetry between quark and lepton numbers. But there is no reason why the number of leptons has to be limited to six either.

In a sense, there are not six quarks, but eighteen. Quarks not only come in six *flavors* (up, down, strange, charmed, bottom and top), but they also come in three different *colors*, designated *blue, green,* and *red.* Naturally, this property of quarks has nothing to do with the colors of ordinary objects. Color is a property of light, which is emitted and absorbed by atoms. Quarks, which are constituents of constituents of atoms, do not reflect or emit light. Thus they cannot have color in the ordinary sense of the term.

The concept of color was first suggested by physicist Oscar W.

Greenberg of the University of Maryland in 1964. Greenberg's idea was originally an ad hoc hypothesis which was invented to explain one of the problems associated with the original quark theory, the existence of hadrons that were made up of three up quarks, three down quarks, or three strange quarks.

The reason why this was a problem was that the Pauli exclusion principle implied that such hadrons should not exist, at least not in the case where all three quarks occupied the same energy state. The exclusion principle states that no two fermions in close association with one another can have the same energy. The existence of spin allows the principle to be circumvented to some extent. Two neighboring fermions can have the same energy if their spins are oriented in opposite directions. But there is no way that a third particle with identical properties can be added, since quark spins have only two possible orientations. Particle C must align its spin either with particle A or with particle B, and the exclusion principle says that this is impossible.

But if the particles possess some additional property that distinguishes them from one another, the exclusion principle does not apply. Three quarks can occupy the same energy state if their colors are different. If there are three different colors, the three identical quarks that make up certain baryons can all have the same energy and the same spin. In such a case, the quarks are three different kinds of particle, at least as far as the exclusion principle is concerned.

The concept of quark color eventually turned out to be more significant than it had originally seemed. During the mid-1970s, theoretical physicists were able to work out a theory of quark interactions that was analogous with quantum electrodynamics. It was the property of color which provided the clue to the forces that bound quarks together.

This theory was called *quantum chromodynamics*, or QCD. In QCD, color is viewed as a property that is analogous to electrical charge. Color charges bind quarks together in hadrons in the same way that electrical charges cause electrons and nuclei to be bound together in atoms.

According to QCD, baryons and mesons are *colorless*. The colors of the quark constituents of a particle always cancel out to produce a net color charge of zero. For example, a proton is thought to contain one blue quark, one green quark, and one red quark. The cancellation of color charge is analogous to the production of white light when these three primary colors are mixed. In the case of mesons the cancellation is even simpler. A meson is a combination of a quark and an antiquark. According to QCD, an antiquark must have an *anticolor*. For example, a red quark may combine with an antiquark that is colored antired. The idea of anticolor, by the way, is not as bizarre as it sounds. It is analogous to the idea of complementary colors of light. Yellow and blue light, for example, combine to produce white.* The color yellow is said to be the complement of blue. In QCD, the terminology is even simpler. Rather than introduce new colors, such as yellow, one simply speaks of antired, antigreen, and antiblue.

Quantum chromodynamics is quite a successful theory. It explains the forces that bind quarks together in hadrons, and it explains the force which binds protons and neutrons together in nuclei. According to QCD, the color force which acts between quarks *is* the strong nuclear force. The force between nucleons is nothing more than a by-product of the color interaction. If mesons are carriers of the force between nucleons, and if mesons are made of quarks, then it is really the interquark forces that cause neutrons and protons to stick to one another.

QCD is a more complicated theory than its analogue, QED. In QED, there are only two kinds of charge, positive and negative. In QCD, there are three (or six, if anticolors are counted separately). One might expect, therefore, that more than one kind of virtual particle might be needed to act as a carrier of the color force. This is indeed the case. QCD requires that there be eight different *gluons*. Gluons are the analogues of the virtual photons which transmit the electromagnetic force, according to QED.

QCD is quite an elegant theory, but it does not have the simplicity that the physicists who proposed the original version of the quark theory were seeking. Three quarks have not proved to be adequate to explain observed phenomena. Six different quark flavors are now known. According to QCD, each of these six

---

* Of course yellow and blue *pigments* combine to form green. However, the addition of colors of light, which is the more fundamental process, works in a somewhat different way.

quarks can have any of three different colors. Thus the number of different kinds of quark has been multiplied to eighteen (or thirty-six, if antiquarks are counted separately). In addition, there are eight gluons, making twenty-six particles in all. It appears that the search for simplicity has only led to a new kind of complexity.

Perhaps quarks are not really the fundamental constituents of matter. There is no reason why quarks could not be made of yet smaller particles. If they are, the same might be true of leptons. The fact that there are six quark flavors and six different kinds of leptons suggests that there is some kind of symmetry between the two particle classes.

Although the idea of quark constituents is not intrinsically unreasonable, there is not yet any experimental evidence to support it. Nor does there seem to be any possibility that this *prequark* hypothesis can be tested in the foreseeable future. Very high energies are required if one wants to penetrate deeply into the structure of matter. The reason is a simple one. According to quantum mechanics, particles of high energy have short wavelengths. One must make use of very short-wavelength particles in order to "see" such tiny objects as quarks; otherwise the picture will be too "fuzzy" to yield any useful information. The energies that can be attained in present-day particle accelerators are simply not sufficient to resolve quarks and leptons into their components—if indeed such components exist. A particle accelerator of dimensions much larger than our solar system might have to be built before one had any hope of penetrating to the next level.

It may be that quarks and leptons are the ultimate constituents of matter. On the other hand, there may be another level, or several of them. Even the possibility that there are an infinite number of levels cannot be ruled out. For all we know, quarks and leptons are composed of smaller particles, which are composed of yet smaller particles, and so on, and so on. Physical reality could conceivably be a kind of cosmic onion, with level below level below level. It may be that the levels never come to an end.

However, in the absence of any experimental evidence for additional levels, these theoretical possibilities can only be pursued to a limited extent. Theory and experiment must act in concert if one wants to call what one is doing "science." Theoretical speculation that leaves experiment too far behind isn't physics any longer; it is metaphysics.

Therefore the search for an ultimate simplicity in nature has followed a different path in recent years. Since the search for fundamental particles has progressed about as far as it can, given the present level of experimental technology, physicists have shifted their attention to a study of the nature of force.

The fundamental forces known to physicists have not proliferated in the manner that the particles have. In fact, just the opposite has happened. Theoretical physicists have found ways to unify seemingly dissimilar forces within single theories. This process of unification began in the nineteenth century, when James Clerk Maxwell discovered a theory of electromagnetism. Not only did Maxwell's theory demonstrate that electricity and magnetism were different facets of the same phenomenon, but it also predicted a new phenomenon, electromagnetic radiation.

This suggests that if ways of unifying the four forces of nature could be found, physicists might not only obtain a theory which explained the fundamental characteristics of force, energy, and matter, but they might also be able to use the theory to predict the existence of previously undiscovered phenomena.

The first step toward the construction of such a theory was taken in 1967 when the American physicist Steven Weinberg and the Pakistani physicist Abdus Salam independently proposed a theory which unified the electromagnetic and weak forces. The Weinberg-Salam theory received a needed boost in 1971 when the Dutch physicist Gerhard 't Hooft showed that the new *electroweak* theory could be renormalized, that the infinities that it produced could be removed in a manner analogous to that in which the infinites of QED were eliminated.

According to the theory, the electroweak force (now there was only one force, rather than two) was mediated by a set of four particles. One of these was the familiar photon. The others were designated by the letters W and Z. Since the W particle could have either positive or negative electric charge, there were two W's. The theory was a spectacular success; all three of the new particles were discovered in 1983.

The hypothesis that the weak and electromagnetic interactions were different aspects of a single force was thus confirmed.

The four forces had been reduced in number to three. The next step was obvious. One might attempt to see if the forces could be unified still further.

In fact, attempts were made to carry this step out long before the electroweak theory received experimental confirmation. In 1973, the American physicists Sheldon Glashow and Howard Georgi published the first *grand unified theory,* or GUT. Glashow and Georgi were attempting to find a way that the strong and electroweak interactions could be merged into a single grand unified force.

Since Glashow and Georgi published their theory, many other grand unified theories have been proposed. No one yet knows which, if any, of them is most likely to be correct. There is not yet sufficient experimental evidence to allow one to decide among the various GUTs.

There does exist evidence which seems to indicate that the so-called *standard model,* a combination of QCD and the electroweak theory, is not entirely adequate. As this chapter is being written, experiments performed on particle accelerators are beginning to turn up new phenomena which these theories do not adequately explain.

On the other hand, there is as yet little experimental evidence that would favor the grand unified theories. These theories make certain predictions which experimental physicists have so far been unable to confirm. For example, the GUTs imply that the proton is not a perfectly stable particle, as had previously been believed. According to these theories, a proton should decay into other particles (a pion and a positron, for example) after about 10^{32} years, on the average. Now 10^{32} years is about 10 billion trillion times greater than the present age of the universe. But this does not imply that proton decay should be unobservable. If one could assemble together 10^{32} protons, it would only be necessary to wait a year before one of them disintegrated. And if one had 365×10^{32} protons, the waiting time would be reduced to a day.

It so happens that there are about 10^{32} protons in 300 tons of matter, so it is not impossible to perform the experiment. Naturally that doesn't imply that it is easy to carry out. Not only must one watch for a very rare event, it is also necessary to eliminate background radiation that might confuse the results. One experiment, for example, was performed in a salt mine under Lake Erie in order to shield the experimental apparatus from cosmic rays.

As I write this, the existence of proton decay has not yet been confirmed. But neither this, nor the fact that no one knows which GUT is correct (or indeed, if any of them is) has prevented theoretical physicists from attempting to go a step further, and to devise a theory that would provide a unified explanation of all four forces.

There is a promising approach to unification that is known as *supergravity*, and a number of supergravity theories have been formulated. But before I explain what supergravity is, it might be a good idea to comment on the reasons why gravity has been the last of the four forces to be included in attempts at unification.

The formulation of a quantum field theory of gravity is an especially difficult task. In fact, there is not even any evidence to confirm that gravity has a quantum character. The graviton has not been detected. Gravity is so weak a force, compared with the other three, that there is a chance that the graviton will never be found. Furthermore, it appears that a quantum field theory of gravity cannot be renormalized; there seems to be no way to get rid of the infinities that crop up, at least not if one assumes that the force of gravity is mediated by only one kind of particle.

Nevertheless there is a chance that unification of all four forces might eventually be successful. The difficulty with the non-renormalizability of gravity can be overcome if one applies a mathematical technique known as *supersymmetry* (which theoretical physicists have nicknamed *susy*). If one makes use of this principle in constructing a theory of gravity, it is found that the graviton is not the only quantum particle associated with gravitational force. Particles called gravitinos must also exist. This solves the problem of renormalization because the infinities produced by the gravitinos cancel out those associated with the gravitons.

However, the gravitino differs from all the force-carrying particles known previously. According to supergravity theory, the gravitino is a fermion; it has a spin of $3/2$. This contradicts the previously accepted idea that all force-carrying particles must be bosons.

However, if one is willing to accept this result, then the theory can be extended to include the other three forces of nature.

When this is done, one obtains a combined explanation of the strong, weak, electromagnetic, and gravitational forces. Of course, supergravity theories must be considered to be very speculative. At present, the only justification for them is the simplicity of the assumptions on which they are based. There is no experimental evidence which would indicate that one or another of the supergravity theories is correct.

As one might expect, supergravity theories predict new phenomena. In particular, they imply that there exist numerous different kinds of undiscovered particles. Supergravity theories blur the distinctions between bosons and fermions, and predict that every fermion has a boson partner. Similarly, for every known boson, there must be an as-yet undiscovered fermion.

These hypothetical particles are known as *sparticles*. Examples include the *squark* (the spin-0 analogue of the spin-½ quark) and the *photino* (the spin-½ version of the spin-1 photon). And of course, the gravitino is a sparticle too. It is the spin-³⁄₂ partner of the spin-2 graviton. Nor does the list end there. Other particles predicted by supergravity theory include the *gluino, zino,* and *wino*.

One might think that supergravity makes matters more complicated, not simpler. But this is not really the case. It predicts the existence of new particles, to be sure. But these particles are only different manifestations of the single unified force. If any of these theories are correct, it is this "superforce" that is the fundamental constituent of physical reality. The numerous predicted particles simply reflect the fact that the superforce can appear to us in many different guises.

The superforce—if it really exists—could provide us with an explanation of all the phenomena that are observed in the quantum world. By blurring the distinction between fermions and bosons, supergravity theory creates a unified description of force and matter. One can no longer separate particles into two different categories according to whether they have integral or half-integral spin, and say that the former are carriers of force, the latter constituents of matter. If one of these theories is correct, then everything that exists—force, matter, and energy—is only one of the superforce's manifestations.

This kind of unification of the laws of physics has a great

appeal. But it doesn't necessarily follow that one or another of the supergravity theories must be correct. Only experiment can decide that question. At the moment, experiment has nothing to say on the subject.

Experiments performed on high-energy particle accelerators are beginning to turn up some new phenomena that the theories which make up the standard model do not seem capable of explaining. But this does not imply that supergravity's predictions will be confirmed. It is possible that the new high-energy phenomena could turn out to be evidence for the existence of squarks, gluinos, or other sparticles. On the other hand, they may have nothing to do with sparticles at all. For example, they could be related somehow to the possibility of a composite structure for quarks and leptons. Finally, it is possible that physicists may find themselves confronted by evidence that indicates that neither of these approaches is the right one. All that one can say at the moment is that physicists are getting some experimental results that are not yet understood.

The lack of any experimental support for supergravity is not the only problem faced by its advocates. There are theoretical difficulties too. A theory that is based on simple ideas is not always so simple when it is worked out in detail. The calculations that are done in supergravity theory are often very long, and difficult to carry out. Furthermore, the simplest supergravity theories are known to be incorrect. Although they predict the existence of numerous different kinds of particles—including some new and exotic varieties—they do not seem to be capable of accounting for all of the quarks and leptons that are already known.

I will have more to say about supergravity in the next chapter. In the last few years, theoretical physicists have been investigating the possibility that there may be more than three dimensions of space. Since supergravity theories can be formulated in any number of dimensions up to eleven (ten of space and one of time), this has led to speculation that our universe might actually be eleven-dimensional.

At the moment, however, it might be best to stop and to attempt to separate fact and speculation. It is important to differentiate between ideas that have experimental confirmation and those which could turn out to be nothing but theoretical fantasies.

The following can be considered to be reasonably well-confirmed ideas:

1. Matter is made up of quarks and leptons. There are at least six leptons and at least six flavors of quark.

2. Quarks come in three different colors also. Color is a kind of charge that is analogous to the electric charge that causes electrons to be bound in atoms.

3. The strong, or color, force is mediated by eight different kinds of gluon.

4. The four forces that are observed in nature have been reduced to three. It has been shown that the weak and electromagnetic interactions are different aspects of the electroweak force.

5. The "gluons" of the electroweak force are the photon, the two W particles, and the Z. All four of these particles have now been detected experimentally.

The following are speculation;

1. Quarks and leptons may have constituents. These constituents could have constituents themselves. The number of levels could be finite, or it could be infinite.

2. It may be possible to unify the electroweak and strong forces in a grand unified theory. One possible confirmation of such a theory would be detection of proton decay.

3. Since the other three forces can be described by quantum field theories, it seems reasonable to assume that gravity has a quantum character too.

4. It may be possible to find a supergravity theory that successfully unifies all four forces. The experimental confirmation of such a theory would imply that everything in the universe—force, energy, and matter—was a manifestation of a single superforce. In other words, the essential physical reality would be a single quantum field that would give rise to all forces and all matter.

Part Two

The Nature of the Universe

4

Space and Time

The fact that ideas about the nature of force and matter have changed is not particularly astonishing. Perhaps one would not have expected them to be altered to so great an extent. However, even the most conservative scientists of the nineteenth century would undoubtably have admitted that some changes were to be expected.

On the other hand, it is unlikely that many nineteenth-century physicists would have expected ideas about the nature of space and time to change. According to the view that was accepted a century ago, space and time were nothing more than an arena in which physical events took place. It was thought that the universe was made up of objects that had precise locations within a fixed, absolute space. Time, it was believed, was something that flowed at an even rate throughout the universe.

These commonsense ideas were upset by the special theory of relativity. According to Einstein's theory, there was no such thing as absolute space. Furthermore, the theory said, the measurement of distance was relative, and it was impossible to determine whether or not two events were simultaneous if they were separated in space. If space and time constituted an arena in which things happened, it was not an arena of fixed dimensions. Nor was it an arena in which one could say that distant events happened at the "same time."

According to the special theory, length measurements are relative. However, it is not only objects that appear to shrink; the length contraction applies to distances too. For example, suppose that a spaceship is traveling past the earth at a velocity that is equal to 87 percent of that of light. To observers on the ship, objects on the earth will appear to contract to 50 percent of the length that would be measured by a "stationary" earthbound observer. Observers on the ship will see the same contraction when they attempt to measure distances. Suppose, for example, that the ship is moving toward a certain star. If astronomers on the earth determine that the star is ten light-years away, their colleagues on the spaceship will conclude that it is located at a distance of only five light-years.

Furthermore, one must conclude that both measurements are correct. According to the special theory of relativity, measurements made in one reference system are as valid as those made in another. Since there is no way of detecting absolute motion, one cannot say that one measurement or the other represents the "real" distance. All that it is possible to say is that the star is five light-years away in one reference system and ten light-years away in the other.

A light-year is, by definition, the distance that light travels in one year. Hence one set of observers will conclude that the light which they see left the star five years ago. The other observers will conclude that it has been traveling for ten years. The two conclusions will be equally correct. If it is the year 1996 when the observations are made, observers on the ship will say that they are seeing light that left the star in 1991, while earthbound observers will calculate that it was emitted in 1986.

This implies that observers on the earth will tend to disagree with those on the ship as to whether or not certain events are simultaneous. For example, suppose that the star is part of a binary system and that it was eclipsed by its companion star as the light was emitted. The astronomers on the earth will conclude that this eclipse is simultaneous with certain terrestrial events which took place in 1986, while the crew of the ship will maintain that the eclipse was simultaneous with events that took place five years later.

This implies that the very concept of simultaneity must be discarded. According to the special theory of relativity it makes no sense to say that spatially separated events are, or are not, simultaneous. Nor can one meaningfully speak of "now" in a distant place. The relativity of time implies that the concept of "now" cannot be extended beyond the place we call "here." If there is no simultaneity, the "now" cannot be universal.

One might think that if distances and time intervals shrink and expand in the manner that special relativity says they do, it would be impossible to do physics at all. But matters are not as bad as they might seem. Within each system of reference, distances and times will remain the same. As long as the ship continues to move with a constant velocity with respect to the earth, the times and distances measured by its crew will remain unaltered. The same is true of the earth. Admittedly, the earth changes direction as it moves around the sun, so the earth is not really in a constant-velocity reference system. However, the earth's velocity is very small compared with that of light. Consequently, relativistic effects are so small that they are unmeasurable. As a result, earthbound observers will always conclude, for example, that the star remains at a distance of ten light-years with respect to the earth.

Furthermore, there are some quantities which always remain constant, according to special relativity. These quantities are the same for every observer. One of them is the velocity of light. Every observer will find that light propagates at a velocity of 300,000 kilometers per second, whatever his state of motion.

The velocity of light is a kind of combination of space and time. Velocity is measured in meters, or kilometers, or feet, or miles per second. Though distances and times may change, their combination, velocity, does not (at least the velocity of light doesn't; other velocities transform in a complicated way). Another quantity that always remains the same is the *space-time interval* between events. If the distance between two events is mathematically combined with their separation in time, one obtains a quantity which is constant for every observer. The space-time interval is a kind of four-dimensional "distance" which always remains the same, even though distances and times may differ.

The net result is that although many quantities are relative, the universe of special relativity has certain unvarying features. In fact, the principle of relativity itself can be thought of as an insistence upon the relevance of invariance. The idea that there is no such thing as absolute motion is equivalent to the notion that the laws of physics must remain the same for every observer. If an observer could tell whether or not he was moving with respect to absolute space, this would not be the case. There would be one set of laws for motionless observers, and others for moving observers who would have to use various kinds of correction factors to account for their motion.

As I have noted previously, solving the equations of physics is often a difficult task. Naturally, physicists desire to make things as easy for themselves as possible. If they didn't, they might never get anywhere when faced with really complicated problems. Hence they are always looking for ways to make their theories mathematically simpler.

Special relativity becomes less complicated if one makes use of a concept that was introduced by the Russo-German mathematician Herman Minkowski in 1907, two years after Einstein's theory was published. This is the concept of four-dimensional *space-time*. Since the constants in relativity tend to be combinations of space and time, it seems natural to introduce a four-dimensional framework at the outset.

It is possible to formulate the equations of relativity without making use of this device. In fact, Einstein did just that in his original paper. However, after he saw what Minkowski had done, he recognized the value of the concept, and spoke of space-time in his later work.

The use of the idea of space-time does not imply that time is a kind of space. Nor does it imply that an extra dimension has to be added to the original three. If special relativity is a "four-dimensional" theory, then so is Newtonian mechanics. Newtonian physics also makes use of one time dimension and three of space. The only difference is that in Newton's theory, distances and times remain constant, and hence there is no advantage to considering combinations of the two.

If Einstein had propounded only his special theory of relativity, the concept of space-time would have no fundamental importance. It would be a useful but unnecessary mathematical technique. After all, Einstein was able to formulate the special theory before the concept was invented. On the other hand, without the idea of space-time, Einstein's general theory of relativity could not have been worked out.

The general theory, which was propounded in 1915, is Einstein's theory of gravity. According to this theory, gravitation is the result of the curvature of space-time. Objects which move in gravitational fields, according to the theory, do not behave the way they do because forces act upon them. On the contrary, they simply follow paths of least resistance in curved space and time.

In order to see what the concept of relativity has to do with gravity, it might be best to take another brief look at the special theory. Special theory describes effects which are seen when observers travel at constant velocities with respect to one another. However, it says nothing about the phenomena which appear when these observers experience accelerations. It holds only when they continue on at the same speed in the same direction.

In the general theory this restriction is dropped. The theory is, in effect, a theory of accelerated motion. One would expect that in such a theory, the time and space coordinates would interact with one another in more complicated ways than they do in the special theory. After all, acceleration is a more complex combination of space and time than velocity. Velocity is measured in miles per hour, or in meters per second, or in some similar system of units. Whatever units are used, space (miles or meters) and time (seconds) appear only once. But time appears twice in expressions of acceleration. For example, if an object is dropped and allowed to fall to the ground, it will accelerate at a rate of 32 feet per second per second.

The "per second" appears twice because acceleration is a change of velocity. Every second that an object is falling, its velocity will be 32 feet per second greater. After one second, it will be falling at a rate of 32 feet per second; after two seconds, its velocity will be 64 feet per second; after three seconds, it will be 96; and, if it has been dropped from high enough an altitude that it is still falling after ten seconds, it will be moving at a rate of 320 feet per second.

If general relativity dealt only with phenomena seen by accelerated observers, it would not have much significance. However, it has been known since the time of Galileo that gravity causes objects to accelerate. Falling bodies move faster and faster as they move toward the ground. Furthermore, if air resistance is negligible, they all fall at the same rate, no matter how heavy or how light they may be. In fact, if a feather and a lead ball are placed in a vacuum, so that air resistance is eliminated, they will strike the bottom of their container together if they are dropped from the top of it at the same time.

Einstein realized that this implied that acceleration and gravity were somehow equivalent. He saw that if he could express this equivalence in mathematical form, he could relate both gravity and acceleration to the structure of space-time and obtain a theory of gravitation.

There is a simple example that can be used to illustrate Einstein's ideas about the equivalence of acceleration and gravity. Suppose that I am given an anesthetic and that I wake up in a room with no windows. Suppose, also, that I feel a normal pull of gravity. Must I conclude that the room is in a building on the surface of the earth? Or is it also possible that the room is a cabin in a spaceship millions of miles from the earth, and that the spaceship is traveling at an acceleration of one g?

The answer to this question is that the latter situation is possible. Furthermore, if I am not allowed to leave the room, there is no experiment that I can perform that would allow me to distinguish one possibility from the other. The effects of the earth's gravitational pull and a one-g acceleration are the same. The use of the term g (for "gravities") as a measure of acceleration is a reflection of this.

The equations of general relativity imply that space-time should appear to be curved to an accelerated observer. But, since there is no way to distinguish gravity from acceleration,* space-

* More precisely, there is no way to distinguish acceleration from a *local* gravitational field, the gravitational field at one particular point in space. There is no acceleration that could mimic the spherically symmetric field of the earth everywhere, for example.

time should also be curved by gravitational masses such as the earth or the sun. It is possible to conclude that the earth revolves around the sun, not because gravitational force acts upon it, but because the space-time through which it moves is distorted.

Some of the ideas that one encounters in contemporary physics are difficult to visualize. As we have seen, it is not easy— perhaps it is impossible—to create a perfectly accurate mental picture of what the motion of an electron in an atom "looks like." At first glance, the idea of curved space-time seems even more difficult.

But perhaps things are not as bad as all that. It is possible to invent analogies which give us some idea of the properties of curved space-time. To begin with, it is possible to speak of the effects of the curvature of space and the "curvature" of space-time separately. The latter is somewhat simpler, so it will be examined first.

The effects that gravitational masses have on time are similar to those of velocity in the special theory of relativity; gravitational fields produce time dilations. Time will flow more slowly in the intense gravitational field that exists at the surface of the sun, for example, than it will at a point in space that is far from any massive object. Gravitational time dilation can even be measured in the relatively weak gravity of the earth.

For example, in 1971, an experiment was performed in which some physicists took accurate atomic clocks aboard commercial jetliners and flew around the world with them. It was found that the decrease in the earth's gravitational pull at high altitudes allowed the clocks to run slightly faster. Later experiments, in which atomic clocks were placed in military aircraft, and in rockets, have confirmed this result. When gravity weakens, clocks run a bit faster; then they run more slowly again in the slightly stronger gravity that exists at the earth's surface.

The time dilation caused by the sun's gravity has been measured also. Experiments have been performed in which radar beams have been bounced off Venus, Mercury, and Mars when these planets have been positioned on the opposite side of the sun from the earth. It has been found that when such a radar pulse grazes the sun, the sun's gravity will cause a slight time delay, just as the general theory of relativity predicts.

The very first experimental test of general relativity confirmed the theory's predictions concerning the curvature of
space. In 1919, a team of scientists headed by the British astronomer Arthur (later Sir Arthur) Eddington went to the island of
Principe, which is located off the coast of Africa, to observe a solar
eclipse. According to Einstein's theory, the curvature of space
near the sun should bend the path of a ray of starlight that grazes
the sun's surface. Since starlight that passes near the sun can be
seen only during total eclipses, it was necessary to mount an expe-
dition to make the necessary observations.

Eddington and his colleagues were able to confirm the predictions of Einstein's theory within the limits of experimental uncertainty. However, the observations were so difficult and the possibility of experimental error so great that the confirmation was not
entirely conclusive. But Eddington's experiment was improved
upon during the 1970s, when astronomers observed the deflection, by the sun, of radio waves emitted by distant astronomical
objects. As experimental techniques have become more sophisti-
cated, the general theory of relativity has been confirmed to an
ever-increasing degree of accuracy.

When we say that space-time is curved, we mean that time
slows down in a gravitational field, and that the path of light or
other radiation will be deflected by gravitational masses. But what
about the orbit of the earth? Does the fact that the earth revolves
around the sun at a distance of approximately 93 million miles in
an elliptical orbit imply that space is curved into an ellipse of
approximately that radius? No. In this case, matters are a bit more
complicated. But fortunately there is a widely used analogy that
makes the effects of spatial curvature upon the motion of the
earth relatively easy to understand.

Imagine that a ball is rolled across a rubber sheet. If the sheet
is stretched flat, the ball will travel in a straight line. This is analogous to the motion of an object that is far away from any gravitating masses. In such a situation, the curvature of space will be close
to zero, and there will be no discernible effects upon the object's
motion.

Now imagine that weights are attached to the underside of the
sheet, creating depressions in it. If the ball rolls near one of these
depressions, its path will be deflected by the curvature of the

rubber sheet. If the ball is not traveling too rapidly, it can be caused to revolve in a circular orbit around the center of one of these depressions in the same way that a ball will spin around a roulette wheel. Obviously, the curvature of "space" (i.e., of the rubber sheet) is not the same as the curvature of the ball's orbit.

The analogy has its limitations. In order to transform the two-dimensional sheet into a "curved space," one must deform it along a third, vertical, dimension. However, when we say that three-dimensional space is curved, we do not mean that it is curved in a fourth dimension. Nor, when one speaks of curved four-dimensional space-time, is any fifth dimension invoked. In general relativity, space and time are curved, but there are no extra dimensions that they "curve in."

This does not mean that there is anything paradoxical about the notion of curved space; it is simply an indication that the analogy has broken down. This breakdown has to do with the fact that a rubber sheet is a physical object, while space is not. When one says that space is curved, it is not implied that space is a "thing" that undergoes deformation. One only means that the geometry of space is altered.

A *flat*, uncurved space is described by the Euclidean geometry that we are taught in high school. Curved space, on the other hand, is described by non-Euclidean geometry. In non-Euclidean geometry, the postulates and theorems that describe geometrical figures are altered somewhat. For example, one of the postulates of Euclidean geometry states that it is possible to draw exactly one parallel to a line through any given point outside the line. In non-Euclidean geometries (there is more than one kind), there may be either no parallel lines or an infinite number of them. In Euclidean geometry, the sum of the angles of a triangle is always 180 degrees. In non-Euclidean geometries, the sum of the angles may be less than 180 degrees, or greater.

The reader who has trouble visualizing the curvature of space should not be troubled. Physicists can't visualize it either. What they can do is work out the mathematical properties of a space that has a geometry that is somewhat different from the geometry they were taught in high school. The non-Euclidean geometry of Einsteinian space is one in which a ray of light will travel from one point to another in the shortest time if it follows a path that appears to be deflected. It is a geometry in which planets will follow elliptical orbits when there are no forces acting upon them because the sun curves the space through which they travel. And, if one considers all four space-time coordinates, it is a geometry in which time sometimes slows down.

Although the general theory of relativity is an extension of the special theory, it is somewhat less relativistic than the latter. In special relativity, there is no such thing as absolute motion. However, in the general theory, space-time is no longer conceived to be a featureless arena in which events take place. Space is something that is filled with gravitational hills and valleys. Hence it is possible to determine whether or not a body is in motion relative to the average distribution of matter in the universe. It is sometimes said that when Einstein formulated the general theory, he reinstated the ether. Naturally this does not mean that Einstein changed his mind about the existence of the substance that nineteenth-century physicists believed in. The "ether" referred to in this context is nothing more than an absolute frame of reference.

It is fortunate that such a preferred frame of reference does exist. If one did not, it would not be possible to speak of the distances between stars and galaxies without ambiguity, or to speak of the time that has passed since the creation of the universe. If time and space were completely relative, it would be difficult, or impossible, to formulate a theory that would describe the structure of the universe as a whole.

In the next chapter, I will discuss the question of the implications of general relativity concerning the nature of the universe, However, at the moment, I want to make a few comments about the range of validity of the theory, and to discuss other speculation about the nature of space and time.

Every theory has its limits. Newtonian mechanics can only be used when one deals with velocities that are small compared with that of light. Special relativity is a more far-ranging theory, but it is incapable of dealing with phenomena seen by observers who do not maintain a constant velocity with respect to one another. Thus it should come as no surprise that under certain extreme conditions, the general theory of relativity should also break down.

In fact, one can describe the conditions under which the theory should cease to be valid in a fairly precise manner. At very small distances and over very short periods of time, quantum effects become quite significant. For example, we have seen that the uncertainty principle implies that virtual particles should come into existence for short periods of time, even when there is not enough energy to create them. One should not be surprised, therefore, to discover that the principle implies that quantum uncertainties should begin to affect the nature of time and space themselves under certain conditions.

No one really knows how space-time would be altered. It is only possible to say that the quantum effects should become important at distances of about 10^{-33} centimeters and at times of 10^{-43} seconds. When one deals with distances and times of this magnitude, the general theory of relativity can no longer be used because the theory depends upon the assumption that space-time is a smooth continuum. When quantum fluctuations begin to affect the nature of space-time in an unknown way, this assumption is no longer valid.

It is possible to calculate a quantity, called the *Planck distance*, below which quantum fluctuations become important. The Planck distance is equal to 1.61×10^{-33} centimeters. In practice, the figure of 10^{-33} centimeters is more commonly used, since there is no precise point at which quantum effects should suddenly begin to appear. They gradually become more and more important as one considers smaller and smaller distances.

If one divides the Planck distance by the velocity of light, one obtains a quantity called the *Planck time*, which is equal to 5.36×10^{-44} seconds. This is just the time that it would take a ray of light to travel 1.61×10^{-33} centimeters. This figure is also normally rounded off, and the Planck time is often given as 10^{-43} seconds.

One can only guess what space and time might be like in the Planck region. It may be that such concepts as "before" and "after" lose their meaning, and that "space" no longer exists in the familiar sense of the term. University of Texas physicist John Archibald Wheeler suggests that space might even take on the character of a churning foam. He theorizes that on the Planck scale, space may be full of tiny holes, bridges, and tunnels that are created for tiny fractions of a second, and which dissolve almost as quickly as they appear. If this kind of ferment exists, there is probably no way for us to detect it. From our point of view, space and time appear to be smooth and featureless, just as the often violent oceans appear to be flat and featureless to one who observes them from a great height.

Recently, there has been a great deal of speculation about the nature of space-time. Some physicists have suggested that space might have more than three dimensions, but that these extra dimensions might be invisible to us for the same reason that we cannot see the quantum fluctuations that presumably take place in the Planck region.

Before we look at some of these speculations, it might be appropriate to ask how we know that there are only four dimensions of space-time at the macroscopic level. How do we know that there is not a fourth, unseen dimension of space that we, as three-dimensional beings, cannot perceive? For that matter, how do we know that there is only one dimension of time?

Perhaps it would be simpler to attempt to dispose of the latter question first. There is probably not any way of proving that extra dimensions of time do not exist. Nevertheless, the possibility is generally not taken very seriously. It seems to be impossible to imagine what an extra dimension of time would be like, and there is no reason to believe that such a thing exists. The idea that one might be able to travel in two "perpendicular" directions of time simultaneously cannot easily be interpreted. It may not even have any meaning. Since there are no good theoretical or experimental reasons for taking such an idea seriously, the possibility is generally ignored.

On the other hand, the idea that extra dimensions of space may exist is one that cannot so easily be ruled out. There does exist an argument which seems to imply that such dimensions do not exist. But, as we shall see, the argument tells us only that if such dimensions are real, they will not be visible on the macroscopic level.

The argument is based on Newton's law of gravitation. Admittedly, this law was superseded by general relativity. But this does not necessarily imply that the argument is invalid. Under most circumstances, Newton's inverse-square law of gravity is an excellent approximation of the equations of the general theory of relativity. Hence its implications have to be taken very seriously.

In most cases, the general theory of relativity needs to be used only when gravitational forces are very intense. Just as the effects predicted by special relativity are insignificant when one considers velocities that are small compared with that of light, the effects predicted by general relativity are small under most circumstances. Newton's law of gravitation describes the motion of the planets in the solar system with near-perfect accuracy. The only exception is the planet Mercury. However, relativistic effects cause Mercury's orbit to be perturbed only to the extent of 43 seconds of arc (about 1/84 of a degree) per century. As a result, any argument that is based on Newton's law must be accepted as valid as long as one cannot find any logical flaw in it.

An inverse-square law of gravitation can be valid if and only if there are three spatial dimensions. In order to see why this should be so, consider an imaginary sphere with the sun at its center. Now, in three-dimensional geometry, the area of a sphere is proportional to the square of the sphere's radius. Thus, if the dimensions of a sphere are doubled, its surface will be increased by a factor of 4. If the dimensions are tripled, the surface area will be 9 times as great, and so on.

According to Newton, gravity was described by an inverse-square law. If the distance from the sun is doubled, the gravitational force that it exerts will be one-fourth as great. If the distance is tripled, the force drops to one-ninth of its former value. It appears that the force of gravity decreases by exactly the same amount that the surface area of an imaginary sphere around the sun increases.

This is no accident. One can see the connection quite clearly if one imagines that the gravitational force exerted by the sun is spread out over a spherical surface. As the area of the sphere increases, the intensity of the sun's gravitational field is correspondingly diluted.

But if there were four dimensions of space, this relationship would not hold. In this case, one would have to imagine that a four-dimensional *hypersphere* surrounded the sun. If the consequences of this are worked out mathematically, one obtains the result that gravity would have to obey an inverse-cube law if space were four-dimensional. That is, if the distance from the sun were doubled, the gravitational force would be only one-eighth as strong (the cube of 2—$2 \times 2 \times 2$—is 8).

But how do we know that an inverse-cube law of gravity is not valid? One could argue that perhaps our inability to perceive the extra spatial dimension causes us to misinterpret the law of gravity too. Perhaps we only think that gravity follows an inverse-square relationship because we cannot see the fourth dimension.

There exist convincing arguments which indicate that this is not the case. It is possible to show mathematically that if gravity did not follow an inverse-square law, then stable planetary orbits would not be possible. In particular, an inverse-cube law would imply that all the planets in the solar system should spiral inward toward the sun. If such a law were valid, the solar system could not have endured for the 5 billion years that have passed since the sun and planets were formed.

Scientists are a perverse lot. When they are told that something cannot exist, they sometimes go ahead and speculate about its existence anyway. When they are presented with what appears to be a convincing argument, they often try to see if the argument has loopholes. Perhaps it is fortunate that they do this. Too much dogmatism would impede scientific advance. One of the reasons that physicists' conceptions of reality have changed so dramatically during the twentieth century is that scientists have been willing to speculate about the "impossible."

At one time, it was thought to be inconceivable that the atom, which was defined to be the smallest component of matter, should have constituents. Some years later, most physicists believed that it was impossible for subatomic objects to exhibit wave and particle characteristics at the same time. The wave and particle descriptions were thought to be mutually contradictory. In the decades that have passed since this time, science has confirmed the validity of one "impossible" idea after another. In fact, modern physics sometimes appears to have been invented by Lewis Carroll's White Queen, who often believed as many as six impossible things before breakfast.

When it is concluded that the "impossible" is true after all, it sometimes appears that physics has become paradoxical. But of course it has not. The apparent paradoxes are nothing more than shifts in scientists' conceptions about the nature of things. If theories sometimes seem bizarre, that is only a result of the fact that the constituents of physical reality are being viewed in new and unfamiliar ways. A theory that contained real paradoxes (logical contradictions) would have to be discarded immediately.

One should not be surprised, therefore, to discover that scientists have been speculating about extra dimensions of space since 1914, when the Finnish physicist Gunnar Nordström proposed a five-dimensional theory in an attempt to formulate a unified explanation of the forces of electromagnetism and gravity.

Nordström's theory, which postulated the existence of four dimensions of space and one of time, eventually had to be discarded because it could not explain the bending of light near the sun that was observed by Eddington in 1919. However, this did not keep physicists from considering the possibility that there might be a fourth dimension of space. In 1919, the Polish physicist Theodor Kaluza formulated a five-dimensional version of general relativity. Like Nordström's theory, Kaluza's also explained gravitational and electromagnetic forces. Furthermore, it had the advantage that it was consistent with Eddington's experimental result. Kaluza's theory was an early attempt to unify the forces. According to Kaluza, if the universe were really five-dimensional, then there could exist a single force which could manifest itself both as electromagnetism and gravity.

In 1919, a submission to a scientific journal could be published only if it was endorsed by a well-known physicist. Hence Kaluza, who was only a *privatdocent* (a kind of assistant professor) at the University of Königsberg, sent his paper on the theory to Einstein. Einstein was impressed by the idea, but he suggested that more work should be done before the paper was published. He told Kaluza that the latter's arguments did not appear to be convincing enough, and that attempts should be made to show how the theory could be confirmed by experiment.

Two years later, after Einstein had become aware of an attempt by the German mathematician Hermann Weyl to unify the theories of gravity and electromagnetism in a somewhat different manner, he changed his mind. "Your approach seems in any case to have more to it than the one by H. Weyl," he wrote to Kaluza. "If you wish I shall present your paper to the academy after all." Einstein endorsed Kaluza's speculations, and the paper duly appeared in the journal *Sitzungsberichte der Berliner Akademie* in 1921.

By the time that Kaluza's theory was published, it was already obvious that it had certain shortcomings. The theory was unable to explain quantum phenomena. Furthermore, Kaluza did not explain why the fifth dimension was not observed. For that matter, it was not clear whether the extra dimension was to be understood as physically real, or only as a mathematical fiction.

An attempt to remedy these defects was made in 1926, when the Swedish physicist Oskar Klein attempted to determine whether or not Kaluza's theory was compatible with quantum mechanics. He found that Kaluza's hypothetical unified force would produce gravitational and electromagnetic waves in the four known directions of space-time. According to quantum mechanics, these waves could be interpreted as particles. It seemed that a quantum version of the theory was possible.

Klein also provided an explanation for the fact that the fifth dimension was not observed. He suggested that we do not see it because it is "rolled up" to very small dimensions. When he calculated these dimensions, he obtained a result of 10^{-32} centimeters, or about 10^{20} times less than the diameter of an atomic nucleus. Although Klein did not consider the problem of why the planets did not spiral into the sun, his hypothesis did provide an answer to the question. If the extra dimension was "rolled up" in this manner, the number of macroscopic dimensions would still be four, and the inverse-square law would hold.

In order to see what such a compacted fifth dimension would look like, imagine that a sheet of paper is rolled into a cylindrical shape, and then rolled up more and more tightly until it begins to resemble a long, thin rod. The sheet of paper will still be two-dimensional, but one of the two dimensions will be compacted to a very small size. A compacted fifth dimension would have similar properties. The only difference could be that the amount of compaction would be much greater than that which could be achieved by rolling up a piece of paper.

Another way of stating the same thing would be to say that the 10^{-32} centimeters is the circumference of the universe along the extra dimension. If any object could travel around such a circle, it would circumnavigate the entire universe, and return to its starting point, even though the distance it had traveled would be many orders of magnitude less than the size of a single subnuclear particle.

It all sounds quite fantastic. Nevertheless, if one assumes that our universe does resemble such a rolled-up cylinder, one finds that electromagnetism, like gravity, can be explained by the curvature of space. One cannot help being impressed by the parsimonious nature of the idea; the theory succeeds in explaining quite a variety of phenomena with just a few assumptions. On the other hand, some questions are raised. In particular, is it really legitimate to attribute reality to a dimension of space that is so compacted that there is little possibility that it could ever be observed?

Observation of such a compacted dimension was not possible in Klein's time, and it is not possible today. Since the resolution of small structures requires particles of high energy, a particle "microscope" that was designed to "see" the extra dimension would have to be powerful indeed. It can be calculated that in order to probe the fifth dimension with energetic particles, one would have to build an accelerator that measured thousands of light-years across. Experimental detection of the extra dimension, if indeed it exists, seems to be a rather impractical goal.

After Klein's theoretical investigations were published, Einstein and Pauli did some further work on the theory. However, attempts to pursue the idea were abandoned when it became apparent that there were not just two forces in nature, but four. The discovery of the strong and weak nuclear forces made it apparent that attempts to explain all the forces in nature by postulating the existence of an extra, compacted dimension were hopeless. There was no way that such a theory could accommodate the two nuclear forces.

The existence of the Kaluza-Klein theory was all but forgotten until the late 1970s, when the development of the GUTs and supergravity theories caused some physicists to wonder if there might not be something to the idea after all. Physicists at the University of Texas, Seoul National University, the University of Chicago, the University of Paris, and the California Institute of Technology began to attempt to include the strong and weak forces within the Kaluza-Klein framework by postulating the existence of additional dimensions.

Investigations of the Kaluza-Klein theory today are generally combined with work on supergravity. Physicists have discovered what appears to be a remarkable mathematical coincidence. They are attempting to discover whether it really is just a coincidence, or whether it contains a clue to the nature of physical reality.

Supergravity theories can be formulated in any number of dimensions up to eleven. In twelve or more dimensions the theory breaks down. Furthermore, only one eleven-dimensional supergravity theory is possible. If a smaller number of dimensions is postulated, there are several different possibilities. The coincidence is that eleven is exactly the number of dimensions that one needs in order to account for all four forces within the framework of a Kaluza-Klein theory.

No one knows whether it is just chance that the mathematical restrictions on supergravity theory lead to the same number of dimensions as the requirement that there be a sufficient number of dimensions to accommodate the four physical forces. However, to some physicists this suggests that there might be good reasons for believing that we live in an eleven-dimensional universe. These physicists have expressed hope that a supergravity theory can be formulated which can explain all the observed particles and all the phenomena that experimental physicists have seen.

An eleven-dimensional universe would be one that contained the usual four macroscopic dimensions of space-time as well as seven extra spatial dimensions that were compacted to loops of 10^{-33} centimeters* or less. The seven extra dimensions are presumably "rolled up" in the same way as the one additional dimension that was postulated by Kaluza and Klein. In an eleven-dimensional theory, 10^{-33} centimeters is the approximate size of the universe in seven different mutually perpendicular directions.

At this point, another remarkable mathematical coincidence is encountered. If one postulates the existence of a quantum field that causes some of the extra dimensions of an eleven-dimensional universe to be rolled up, then it turns out that there are only two possibilities. Either four or seven dimensions must be compacted. It is mathematically impossible to roll up dimensions in any other combination.

* Today, the estimate of 10^{-33} centimeters has replaced Klein's 10^{-32}.

If seven dimensions curl up, one is left with the four familiar dimensions of space-time. Of course this is only one of two possibilities. The other would correspond to a seven-dimensional world. This, incidentally, raises the question of whether there is some reason why nature prefers four dimensions to seven, or whether there is a possibility that seven-dimensional space might exist elsewhere in the universe.

Some physicists have expressed the hope that an eleven-dimensional supergravity theory might explain "everything." They think that such a theory might not only provide a unified explanation of all four forces, and the behavior of all observed particles, but might also explain why each particle possesses a particular mass, why the forces have the strengths that they do, and why particles have electric charges of a particular size. In the view of these scientists, the development of such a theory would mark the end of theoretical physics. There would be no more fundamental discoveries to be made; it would only remain to work out some details, and to apply the theory to particular cases.

Other scientists are not so sure about any of this. They point out that it may turn out to be impossible to find an eleven-dimensional supergravity theory that can be shown to be correct. They point out, also, that the detailed studies of eleven-dimensional theory that have been performed so far seem to predict a four-dimensional world that differs from the one we know in certain important respects. There seem to be three problems. First, the eleven-dimensional theory appears to predict the existence of two different classes of neutrinos. One of these kinds of neutrino spins in a manner that is not observed. Second, the theory predicts that the four uncompacted space-time dimensions should be highly curved. But astronomical observations indicate that the average curvature of space in the universe is very small. Finally, eleven-dimensional supergravity theory suffers from the same problem with infinities that plague all quantum field theories. But in supergravity the problem is even worse than it normally is. Ways to remove the infinities have not been found.

Because there are so many unsolved theoretical problems associated with supergravity, some theoretical physicists have begun

to concentrate on investigations of *superstring* theories instead.*
Superstrings are entities in ten dimensions that behave like ordinary particles when the ten dimensions are reduced to four. The
reason that superstring theories are ten-dimensional is that ten or
twenty-six or 506 dimensions must be postulated if the theories
are to be mathematically consistent. Naturally the assumption of
ten dimensions is the simplest of the three possibilities.

Although the problem of how the extra six dimensions might
be compacted has not yet been solved as I write this, some interesting implications of these theories have been worked out. For
example, there are some superstring theories that suggest that
there might be a kind of matter which interacts with ordinary
matter only through the gravitational interaction. Since the production of light and the existence of intermolecular forces are
products of the electromagnetic interaction, not only would this
shadow matter be invisible to us, but one could walk right through
it, no matter how dense it might be. In fact, shadow matter could
be detected only if it was present in large enough quantities to
exert a significant gravitational force on ordinary objects.

The possibility that shadow matter might exist conjures up a
number of science-fiction-like fantasies. One could imagine, for
example, the existence of a shadow earth populated with beings
like ourselves. We would be aware neither of the shadow earth
nor of its inhabitants. In fact, one could walk right through a
shadow mountain and never know it.

But of course all this is fantasy. If shadow matter exists, there
cannot be large quantities of it in the solar system. If there were,
its gravitational effects on the earth and on the other planets
would be noticeable. If shadow matter exists elsewhere, there is
no good reason to believe that the shadow matter universe would
necessarily be populated with shadow beings. The properties of
shadow matter would not have to be the same as those of ordinary
matter. If they were not, the shadow world might not be hospitable to life.

* Fashions in theoretical physics can change rapidly. When I wrote this chapter,
supergravity theories occupied the center of attention. By the time the book was
being edited, interest in superstring theories had increased, and they were in the
spotlight. And of course, new theoretical results could easily shift attention back
to supergravity theories by the time this book is published.

On the other hand, Nemesis, the hypothetical "death star" that supposedly brings destruction to the earth once every 28 million years by causing a rain of comets to fall into the solar system, might be made of shadow matter. There is not yet any evidence that Nemesis exists. However, paleontologists have long known that there have been numerous occasions when mass extinctions have taken place on the earth. One of the theories which attempts to account for these extinctions postulates that the sun has a dark companion star which passes through a cloud of comets every time that it orbits the sun. If this companion star, called Nemesis, were as far from the sun as it is believed to be, it would take 28 million years to complete an orbit. Hence it would be able to perturb the orbits of the comets and send them toward the earth only at long intervals.

Nemesis may not exist. If it does, the fact that it is a dark object would not necessarily imply that it is composed of shadow matter. It could simply be a small star that was too dim to be easily observed from the earth. On the other hand, there is no reason why entire galaxies could not be made of shadow matter. If they were, they would be as invisible as any other shadow matter object would be, and we would have no way of detecting them except by trying to observe their gravitational influences upon ordinary-matter objects.

It must be emphasized that this is all very speculative. There is not yet any experimental evidence to indicate that anything like shadow matter really exists. The idea that our universe might have more than four space-time dimensions has not been experimentally confirmed either. All that one can say is that there is a theoretical possibility that we live in a multidimensional universe. And if we do, we cannot be sure that the number of dimensions is ten or eleven. Theories that postulate the existence of more than eleven dimensions have been investigated too, and it is conceivable that one of them could eventually turn out to be correct. If this happens, the number of dimensions might be large indeed. For example, there is the 506-dimensional superstring theory mentioned previously.

If any of these additional dimensions of space-time really do exist, we may never be able to observe them. On the other hand, the theories could possibly be confirmed in other ways. If the theoretical problems associated with them are solved, it might be possible to develop predictions that could be tested by experiment. For example, it has been suggested that supergravity theory might imply the existence of particles called *graviphotons* which could create repulsive antigravity forces under certain circumstances. The existence of such forces could possibly be verified in the laboratory.

But there is no guarantee that testable theoretical predictions of this sort will be developed. It may be that theoretical physics is beginning to leave experiment behind. If this is indeed the case, questions are raised about the validity of some contemporary theoretical speculation. If physicists propose hypotheses that cannot be verified, are they still doing physics? Or are they engaging in a kind of metaphysical speculation?

Few theoretical discoveries would be made if scientists were not willing to speculate. On the other hand, it is experiment which makes speculation meaningful. Einstein emphasized this point in a letter to Schrödinger in 1950, when he wrote, "Most of [the contemporary physicists] simply do not see what sort of risky game they are playing with reality—reality as independent of what is experimentally established." Einstein himself was always careful to suggest experiments that could be performed to test his theories. As we have seen, one of his reasons for suggesting that Kaluza delay publication of his theory of a fifth dimension was that he did not see how the idea could be experimentally tested.

It is not my intention to single out those physicists who work with multidimensional theories for criticism. I only want to point out that there seems to be a widening gap between theory and experiment. It appears that physicists have advanced the frontiers of science so far that they now run the risk of putting forth questions that are unverifiable, and therefore possibly meaningless.

On the other hand, it is possible to adopt the point of view that even the most far-ranging speculation serves a legitimate scientific purpose. If scientists do not try to imagine what might be true, they will never be able to tell what is true. One could argue that even if we can never determine whether extra dimensions of space exist or not, the discovery that they *might* exist has added to our fund of knowledge. Similarly, if physicists do not attempt to work out the implications of the possible existence of shadow matter we will never know whether the reality of this odd substance is a possibility or not.

Perhaps the wisest course would be to suspend judgment and to observe simply that the nature of theoretical speculation has changed. In the early decades of the twentieth century, theoretical speculation had the same "bizarre" character that it possesses today. However, in those days, speculative ideas were either quickly verified in the laboratory or discarded because they had been shown to be untenable. That no longer seems to be the case today. By the 1980s the trend was to heap speculation upon speculation. It may be too early to tell where all this will lead.

5

The Nature of the Universe

When Copernicus proposed his heliocentric theory in 1543, there were a number of facts that could be adduced against the idea that the sun, and not the earth, was the center of the solar system. It was argued that if it was the earth, and not the sun, that moved, then flying birds would be left behind as the earth rotated under them. Another argument had to do with the observed absence of *stellar parallax*. It seemed that if the earth really did revolve around the sun, then the stars should appear to shift their positions as our planet moved from one end of its orbit to the other.

It was Galileo who pointed out the fallacies in these arguments. His principle of inertia explained why the earth's atmosphere, and the birds that flew in it, were not swept away to the west. The atmosphere would go on moving with the rotating earth as long as there were no forces that would stop it from doing so. The principle was the same as that which caused a falling object to move along with a ship, so that it seemed to drop to the deck in a vertical line.

Galileo explained the inability of astronomers to detect shifts in the positions of stars by suggesting that the stars were much farther away than had previously been thought. In his opinion, the "fixed" stars did have a relative motion with respect to the earth, but the universe was so large that this motion was unobservable. Stellar parallax was not seen because the stars were too distant.

Galileo's argument can be illustrated by means of an analogy. Suppose that one walks down a hallway. In such a case, the relative motion of the walls is easily visible. On the other hand, a distant mountain will not seem to shift position to someone who is walking down the street.

Galileo's argument implied that the universe was much vaster than anyone had suspected. Until Galileo's time, it had generally been thought that the stars were not much farther away from the earth than the planets. The argument about the absence of stellar parallax changed all that. By the middle of the seventeenth century, by which time the Copernican theory was generally accepted, it was realized that the universe was vast indeed.

Some seventeenth-century scientists and philosophers, such as Newton and the French philosopher René Descartes, went so far as to suggest that the universe might be infinite. In fact, Newton believed that he could demonstrate that this must be so. If the universe were finite, he argued, then the force of gravity would cause all of the bodies that it contained to collect together in the center of the universe. In an infinite universe, on the other hand, there would be no center, so this could not happen.

However, the astronomers of succeeding generations recoiled a bit from the idea of infinite space. They tended to view the universe in a more conservative way. By the beginning of the twentieth century, most were convinced that our own Milky Way galaxy *was* the universe. Although the eighteenth-century British astronomer William Herschel and his contemporary, the philosopher Immanuel Kant, had suggested that some of the diffuse patches of light, or *nebulae*, that could be observed through telescopes were other "island universes," astronomers generally tended to discount this idea. In their view, the nebulae were only glowing clouds of gas within the Milky Way.

Today it is known that many of the nebulae are indeed galaxies, and that these galaxies are quite numerous. It is estimated that there are as many galaxies in the observable universe, 100 billion, as there are stars within the Milky Way. However, in 1900, there was no observable evidence that would have indicated that anything like this was the case.

Since 1838, it had been possible to measure the parallax exhibited by certain stars, and to determine their distance from the earth. If the apparent positions of a star were measured when the earth was on opposite sides of its orbit, the distance could be found by triangulation. However, this method worked only for stars that were within about a hundred light-years of the earth. It could not be used to determine the positions of the more distant stars in the Milky Way, which has a diameter of about 100,000 light-years. Attempting to use measurements of parallax to find the distances of the nebulae would have been a hopeless task. Since the telescopes of the day were not powerful enough to resolve individual stars within any of the nebulae, the gas-cloud theory continued to hold sway.

In 1923, the American astronomer Edwin Hubble enlarged the universe in a single stroke. Observing the night sky through the new 100-inch telescope at Mt. Wilson in California, Hubble was able to make out individual stars in the great galaxy in Andromeda, thus confirming that it could not be a nearby gas cloud. Furthermore, by observing certain kinds of variable stars within the Andromeda galaxy, and measuring their apparent brightness, Hubble was able to estimate the galaxy's distance. He concluded that Andromeda was some 800,000 light-years away (this was an underestimate; the true distance is about 2 million light-years). Since Andromeda was a near neighbor of the Milky Way, it was apparent that other, less prominent galaxies must be distant indeed. Hubble was able to estimate distances for galaxies that were 10 times farther away than Andromeda. He noted, also, that there were numerous, similar-looking nebulae that appeared to be many times farther away than that.

Today, we often describe extremely large quantities as "astronomical." Most of us are aware that we live in a cosmos that is so large that the earth and the solar system are only minute specks by comparison. However, the idea that the universe extends for unimaginably vast distances is a relatively modern one. In pre-

Copernican times, there was little reason to think that its size was so great as to dwarf us into insignificance. Even at the beginning of the twentieth century, astronomers had not grasped just how large the universe was.

To be sure, Newton and Descartes had speculated about an infinite universe. But their ideas were not backed up by any observational evidence. It is one thing to speculate about infinity. It is another to actually look out into space and find that one can observe galaxies that are millions of billions of light-years away.

When Hubble showed that distances in the universe were indeed "astronomical," many of his contemporaries found the discovery astonishing. However, this was to be only the first of a succession of discoveries that have changed our conception of the universe. As we shall see, ideas about the nature of the cosmos have been transformed even more dramatically than those about the nature of force, matter and subatomic particles. The scientific conception of the universe has changed so much that about the only thing that has remained the same is the word itself.

The lay person who looks up at the night sky today does not see the heavens in the same way that those who lived in the pre-Copernican era did. They looked up from an earth that was situated in the center of creation, and at stars that were relatively small objects that were not much more distant than the planets. When we look at the heavens, we know that we are seeing innumerable suns like our own, and numerous other galaxies as well.

But a scientist who looks at the heavens sees something entirely different. He knows that the stars and the galaxies are only an insignificant part of the universe, that most of the matter in the cosmos is dark, invisible, and of an undetermined nature. He knows that the universe may have come from nothing, and that it may return to nothing at some time billions of years in the future. He is aware that the universe is in a state of rapid expansion, that distant galaxies are receding from us at enormous velocities. He knows that there is speculation that there might be other universes, perhaps an infinity of them. Finally, he may find himself wondering how it is that the universe happened to be constructed in a strange, improbable way which makes it hospitable to that thing we call "life."

The modern attempt to understand the nature of the universe

begins with Hubble, and with Einstein. Shortly after Einstein proposed his general theory of relativity in 1915, he began to look for solutions to his equations that would describe the structure of the universe as a whole. In 1917, he published a paper in which he described his conception of the cosmos. In this paper, Einstein made what appeared at the time to be a perfectly reasonable assumption. Yet this assumption about the nature of the universe was to lead to what he would later describe as "the greatest blunder of my life."

Einstein began by assuming that the universe was *closed*, that it contained enough matter that space would close upon itself. A closed universe has a finite volume. Its geometry is roughly analogous to that of the surface of a sphere, except that the universe has one more dimension than a spherical surface. Just as a ship or an airplane can circumnavigate the earth, a ray of light could circumnavigate the universe. Such a ray of light would eventually come back to its starting point without ever changing direction (provided, of course, that the universe lasts long enough for the light ray to get all the way around).

Although a closed universe is finite, it has no boundaries. In this respect, it is again analogous to the surface of a sphere. It is difficult, or impossible, to form a mental picture of such a universe. However, this does not imply that there is anything paradoxical about the idea. A closed universe can be described in an unambiguous manner by the equations of general relativity.

Einstein's "blunder" was not his making the assumption that the universe was closed. Although no one really knows whether it is or not, many contemporary physicists also find the idea of a finite universe to be philosophically preferable to that of a universe which is infinite and *open*. Scientists have not yet determined whether the universe contains enough matter to cause space to curve back upon itself, so the question is still undecided. In such a case, one is free to assume whatever one wants in order to see where one's speculation will lead.

If space in a closed universe curves in a way that is analogous to the curvature of a sphere, the curvature of space in an open universe can be compared to that of a saddle. It is not hard to see that a saddle-shaped surface would extend out to infinity if it were not cut off at some point. A saddle curves upward in the front-back direction, and downward from side to side. One can easily imagine a saddle that extends outward in both directions in such a way that the ends never meet.

There is another analogy that might be of some help to those who find infinite saddles hard to picture. Imagine a trombone of infinite length. As the length of the instrument is extended, its mouth flares out to a greater and greater degree. If one ignores the fact that the trombone has a circular cross section (it curves back upon itself in the other direction), such an instrument might be compared to an open universe too.

Einstein's mistake was not his conception of the overall structure of the universe, but his second assumption, that the universe was static and unchanging. Naturally, Einstein had no way of knowing that the universe was expanding rapidly. In 1917, the idea of a static universe had never been questioned. At the time, the idea seemed a perfectly reasonable one. After all, man had been observing the stars since the time of the ancient Babylonians, and the constellations had never been observed to change.

As a matter of fact, constellations do change. If one could travel a million years into the future and look at the sky, one would find them unrecognizable. However, human life—and the lifetime of civilizations—is so short that the changes are not noticeable. The universe, which is "astronomically large," changes on time scales that dwarf those to which we are accustomed. We measure our lives in decades. The life of the universe is measured in tens of billions of years.

As Einstein was attempting to work out the implications of his theory, he discovered that the assumption of a static universe didn't seem to work. His equations implied that the universe had to be in a state of either expansion or contraction. Since Einstein considered this result to be unreasonable, he added a quantity to his equations which he named the *cosmological constant.*

The cosmological constant represented a repulsive force which canceled out the effects of gravity at large distances. It was necessary to assume the existence of such a force because the long-range effects of gravity had to be eliminated somehow if the universe was to remain constant in size. It appeared, however, that the cosmological force was quite an odd one; unlike any of the other forces known to physics, it grew stronger, not weaker, with increasing distance.

If Einstein had any doubts about the existence of such a strange force, he apparently wasn't bothered. Perhaps this was not so unreasonable either. After all, by 1917, Einstein had already made successful use of concepts which appeared, at first, to be even stranger. It must have seemed to him that the idea of a cosmological force wasn't hard to swallow, compared with some of the other ideas that he had proposed. In any case, he went ahead and published his theory of the universe. According to Einstein, the universe was static and finite. It closed upon itself along the spatial dimensions, while time extended into the infinite past and the infinite future.

Unfortunately, the results were not valid, for Einstein had made a second mistake, one that high school freshmen are warned against when they take a course in algebra for the first time. In solving his equations, Einstein had, at one point, divided by a quantity that was sometimes equal to zero. In 1922, this mistake was corrected by the Russian mathematician Alexander Friedmann. Friedmann found that Einstein's universe wasn't static after all. Or rather, it was only quasi-static; small perturbations would cause it to go over into a state of expansion or contraction whenever its equilibrium was disturbed. Such a universe was analogous to an unsharpened pencil that had been stood upright; the slightest disturbance would cause it to go one way or the other.

By the time that Friedmann obtained this result, Einstein had already begun to have doubts about his theory. Although he did not repudiate the idea of a cosmological constant altogether until 1931, he had already commented in 1920 that his constant was "a complication of the theory, which seriously reduces its logical simplicity." It is a pity that he did not discard the idea of a cosmological constant at this point. If he had, he could have predicted the expansion of the universe long before Hubble deduced it from astronomical observations.

Hubble's discovery of the expanding universe, announced in 1929, was based on observations of variable stars like the ones that he had observed in the Andromeda galaxy. In 1912, Henrietta

Leavitt, a Harvard College astronomer working in South Africa, had discovered that there was a relationship between the periods and luminosities of certain types of stars, called *Cepheid variables*. It was her work that made Hubble's discoveries possible.

A variable star is one that brightens, dims, and then brightens again. The changes in brightness are caused by cycles of expansion and contraction. The *period* of such a star is just the time that elapses between peaks of brightness. Leavitt discovered that the brighter a Cepheid variable was at its peak, the longer was the period between peaks. Furthermore, the two quantities were related in a precise way. If a Cepheid had a certain period, then it would always have the same peak luminosity.

This implied that Cepheids could be used to measure distances in the universe. The method was really very simple. First, one determined the period of one of these variable stars. Next, its peak luminosity was computed. Finally, its apparent brightness was measured. This was compared with its intrinsic brightness, and its distance was computed.

The nature of the method can be made a little clearer by means of the following analogy. Suppose that one wants to measure the distance of a light bulb and that, for some reason, this cannot be done with a yardstick or a tape measure. The distance can be determined if one measures the brightness of the bulb. The farther away it is, the dimmer it will appear to be. But if this method is to work, one must know beforehand whether it is a 25- or 60- or 100-watt bulb. A 25-watt bulb and a 100-watt bulb that is twice* as far away will appear equally bright. In other words, one must have a method of determining its intrinsic brightness.

Observations of Cepheid variables allow one to compute distances only of the nearest galaxies. But once those distances are known, other methods can be used to find those of galaxies that are farther away. For example, one can find the ratio of the distances of two galaxies by comparing the apparent brightness of their most luminous stars, or of glowing clouds of gas within the galaxies.

---

* Not 4 times as far, because brightness decreases according to an inverse-square law.

Knowing the distance to galaxies does not immediately allow us to deduce anything about the expansion of the universe. After all, one cannot measure distances, wait thousands or millions of years, and then measure the distances again to see if anything has changed. However, there is a method of determining the rate at which distances are changing. This method is based on measurements of *redshifts*.

When an object is moving toward an observer, the wavelengths of the light that it emits will be shifted toward the blue end of the spectrum. When it is moving away, its light will be shifted toward the red. The existence of this *Doppler shift* is not a relativistic effect. It is simply a consequence of the fact that light is a wave phenomenon. The same efect is observed in the case of sound, which is also a pattern of waves. Sound emitted by an object that is moving away will have a lower pitch, and sound emitted by an object that is moving toward an observer will have a pitch that is higher.

Consider the case of an object that is moving away. As it moves, it emits a light wave that is made up of a series of crests and troughs that are analogous to the crests and troughs of an ocean wave. As the object recedes, each successive wave crest must travel a longer distance before it reaches the observer. This has the effect of spacing out the crests and making the wavelength longer. If, on the other hand, the object is traveling toward the observer, the crests will be bunched together, and the wavelength (defined to be the distance from crest to crest) will be shortened.

The wavelengths of light are correlated with color. Red light has a relatively long wavelength, while that of blue light is shorter. If wavelengths become longer when an object recedes, the light that it emits will be reddened.

Light from stars or galaxies that are moving away from us does not really look redder to the eye, for these objects also emit ultraviolet "light." As the visible part of the spectrum is reddened, some of the normally invisible ultraviolet wavelengths shift into the visible part of the spectrum.

As far as the eye is concerned, nothing has happened. However, when one examines the light with scientific instruments, things have a somewhat different appearance. It is possible to pick out certain characteristic wavelengths, and to determine just how far they have shifted. Once this is done, it is a simple matter to compute the velocity of recession.

When Hubble measured the velocities of the galaxies whose distance he had measured, he found that most of them were moving rapidly away from the earth.* Furthermore, the more distant a galaxy was, the greater the velocity of its recession. Hubble realized that this could mean only one thing. The universe was expanding.

In order to see why this result would indicate a universal expansion, consider the following, often-used analogy. Suppose that a lump of raisin bread dough is placed in an oven. As the dough expands, all of the raisins will move away from one another. Furthermore, the farther apart two raisins are to begin with, the greater their velocity of mutual recession will be.

Hubble's measurements of the rate of expansion allowed him to estimate an age for the universe. Hubble reasoned that if the universe was expanding, it must once have been in a very compressed state. It was therefore a simple matter to compute how long the expansion had been going on. Incidentally, one does not have to know whether the universe is infinite or finite in order to carry out this calculation. An infinite universe naturally does not grow any larger as it expands. But the galaxies contained in it do move farther apart. Thus if one assumes that the expansion has been going on at a fairly constant rate, one finds that the galaxies must once have been very close together. The calculation is exactly the same as it is for a finite universe.

Hubble's estimate for the age of the universe was 2 billion years. We now know that this result was far too small. It seems that there were certain systematic errors in Hubble's distance scales. These errors did not have any effect on his conclusions concerning the existence of an expansion, but they did affect his estimate of the time that the expansion had been going on. Nowadays, many astronomers think that the universe is approximately 15 billion years old. There is some dispute about the exact figure;

---

* The exceptions were certain nearby galaxies. Some of these exhibited blueshifts, which indicated that they were approaching. Today we know that these galaxies are members of a small cluster known as the *local group*. The galaxies of the local group are gravitationally bound to one another.

there are some who would raise or lower it by about 25 percent. However, all agree that the figure of 15 billion represents the right order of magnitude.

The problem of determining whether the universe is open or closed is much more difficult than that of estimating its age. Even if the universe is spatially finite, it is so large that it is impossible to obtain an estimate of its volume from astronomical observations. Astronomers can see galaxies that are more than 10 billion light-years away, but observation of these galaxies does not provide them with any hints as to how much space might lie beyond them.

Attempts to draw some conclusions about the nature of the universe in which we live have generally been based on measurements of the amount of matter that it contains. According to the general theory of relativity, there is a relationship between the matter content of the universe and its "size." The average density of matter in an open, infinite universe is less than that in one that is closed and finite. It has been calculated that if the average matter density is less than about 5×10^{-27} kilograms per cubic meter (about a twentieth of an ounce per million billion cubic miles, or approximately three hydrogen atoms per cubic yard), then the universe is open. If the density is greater than this critical value, then the universe contains enough mass to cause space to curve back upon itself, and the universe is closed.

Three hydrogen atoms per cubic yard does not sound like very much. However, it is about 10 times the amount that is observed to exist in stars and galaxies. The distances between stars and galaxies are so great that the matter density is not very high when an average is taken over all space. A star is an extremely dense object, but the volume of nearly empty space that surrounds it is great indeed.

The observation that the mass contained in stars and galaxies is much less than the critical density does not imply that the universe must be open. As we shall see, there are other kinds of matter in the universe, and it is not an easy task to determine how much of it there is.

The first attempts to measure the matter density of the universe were carried out in a somewhat roundabout way. They were based on the idea that the more mass there is in the universe, the more rapidly the expansion should be slowing down.

The expansion of the universe is retarded by the mutual gravitational attraction of the matter that it contains. The more gravitational mass there is, the more rapidly the expansion will slow. In fact, in a closed universe, the matter density is so great that the expansion must eventually come to a stop. If this happens, a phase of contraction will eventually set in, and the universe will become compressed into an extremely small volume. If the universe is open, this will never happen. Although the expansion will slow, it will never stop completely, and the galaxies will continue to recede from one another forever.

In principle, it should be possible to measure the slowing of the expansion. After all, if one looks 10 billion light years into space, one is also looking 10 billion years into the past. This follows directly from the definition of the term "light-year." If an object is a light-year away, we observe it by the light that it emitted a year ago.

It appears that all that astronomers would have to do would be to observe distant galaxies, measure the rate at which they were receding from one another billions of years ago, and compare this with the rate of recession that is observed today. In practice, it is difficult to carry such observations out. The uncertainties are so great that it is hard to tell whether the expansion of the universe has slowed by any significant amount.

The problem is the uncertainty associated with distance measurements. If one does not know how far away a galaxy is, one cannot determine how long ago it emitted the light by which it is being seen. If one cannot determine this, one does not know how far one is looking into the past. And distance measurements are uncertain indeed. Even the most luminous supergiant stars cannot be detected at distances of more than 30 million light years. As a result, the only way to estimate the distances of faraway galaxies is to use the galaxies themselves as distance indicators.

Astronomers estimate such large distances by observing the most luminous galaxy in a cluster. They assume that it has a certain characteristic brightness, and compare this with the galaxy's apparent luminosity. This gives the approximate distance. The accuracy of this method depends upon two assumptions. The first of these, that the brightest galaxy in any cluster will always have about the same luminosity, is probably a fairly reasonable one. In any case, there are so many clusters of galaxies in the universe that uncertainties can be eliminated to a great extent if one makes use of statistical averages.

The other assumption, that the brightness of a galaxy will remain relatively constant over a period of billions of years, seems to be rather questionable, however. No one really knows how galactic luminosity varies over time. It is not even possible to make the assumption that galaxies have grown dimmer as some of the stars in them have grown old and burned out. Large galaxies— and it is the largest galaxies that must be used when distance measurements are made—may become brighter with time. It is now known that very large galaxies have a tendency to cannibalize smaller ones. Gravitational attraction causes them to gobble up some of their smaller neighbors, becoming brighter and more massive in the process.

The accuracy with which the distances of faraway galaxies can be measured is good enough for some purposes. In fact, it is greater than the accuracy with which the age of the universe can be stated. But this is not good enough. If one wants to know how the present expansion rate compares with expansion rates in the past, a great deal of precision is required. This precision has not been attained.

The other method of measuring the mass density of the universe is a more direct one; it is based on observations of celestial objects, primarily galaxies. If the brightness of a galaxy is measured, one can estimate the number of stars that it contains. If one knows the mass of an average star, the mass of the galaxy can be calculated. Since this method is relatively straightforward, one might think that it would give more accurate results than the one described previously. But, unfortunately, it doesn't.

Measurements of the brightness of a galaxy will only tell us how much matter is contained in its stars, and in other luminous matter, such as glowing clouds of interstellar gas. They do not allow us to estimate the amount of dark matter that might be present in a galaxy. Astronomical measurements are based on the observation and analysis of light and of other forms of radiation that come from the sky. If something emits no radiation, it is not easy to tell how much of it is present.

Since the early 1930s, it has been known that clusters of galaxies contain a great deal of dark matter. This fact was first pointed out in 1933 by the California Institute of Technology astronomer Fritz Zwicky. While studying a large cluster of galaxies in the constellation Coma Berenices that was apparently bound together by the galaxies' mutual gravitational attraction, Zwicky discovered that the amount of mass present in the galaxies' stars was only a fraction of the amount needed to hold the cluster together. According to Zwicky, there seemed to be a *missing mass* problem.

The term "missing mass," which is still frequently used today, is a rather misleading one. "Hidden mass" would really be more accurate. After all, Zwicky did not discover that anything was missing from the Coma Berenices cluster; he simply concluded that the cluster contained matter that he could not see.

A considerable amount of hidden mass exists in all clusters, including the local group. The quantity of hidden mass that they contain is generally at least 10 times the mass that can be observed in stars. Since it is possible that even more hidden mass could exist in the vast spaces between clusters, and since there is no way that the presence of this dark matter can be detected, the overall mass density of the universe is still unknown. According to the best estimates, it is probably between $1/10$ and 10 times the critical amount. The assumption that we live in a closed universe and the assumption that it is open are both consistent with current observations.

One might think that the hidden mass could be made up of interstellar and intergalactic gas and dust. However, observations indicate that this is not the case. In fact, when one speaks of "visible matter," gas and dust have already been taken into account. Gas and dust are easily detected. Gas emits radiation that can be observed from the earth. For example, cool hydrogen gas in interstellar space emits radio waves at a wavelength of 21 centimeters. Gas that has been heated to high temperatures by the radiation emitted from stars emits X rays. It is possible to estimate the amount of gas that is present in a given region of space by measuring the intensity of radiation coming from that region. It is estimated that the mass of the gas that is contained in galactic clusters is roughly equal to the amount that is present in the galaxies' stars.

Interstellar dust is even easier to observe. It acts as a kind of "cosmic smog" that partially or completely obscures the stars behind it. If intergalactic space contained enough dust particles to contribute significantly to the mass of the universe, distant galaxies would appear as hazy as buildings in a city with polluted air. Since this is obviously not the case, one can conclude that the hidden mass is not made up of dust either.

During the early 1970s, it was discovered that some of the hidden mass exists within and just outside the galaxies. Galaxies, including our own, tend to be surrounded by haloes of dark matter. A smaller amount of this dark matter has been found to be present within the luminous inner regions of the galaxies.

In order to understand how these haloes were discovered, it is necessary to know something about the techniques that are used to measure a galactic mass. One method has been described previously. The mass of the luminous matter within a galaxy can be obtained from measurements of the galaxy's luminosity. However, it is obvious that if one wants to know the combined mass of both the bright and dark kinds of matter, a different method must be used.

In order to explain how this method works, I will consider a simpler case first. Imagine that astronomers want to determine the mass of the sun. How would they go about doing this? They would simply observe the orbital motion of the earth, or of any other planet. The velocity with which a planet revolves around the sun does not depend upon the planet's mass; it depends only upon the mass of the sun. If a large, massive planet such as Jupiter occupied the earth's orbit, it would revolve around the sun in exactly the same time that the earth does: one year.

When a planet revolves around the sun, its centrifugal force must exactly balance the sun's gravitational attraction. The more massive a star, the faster a planet must revolve if it is to maintain itself in its orbit. One can calculate the sun's mass from the distance and period of revolution of any planet. Similarly, the earth's mass can be computed from data concerning the orbital motion of the moon.

Stars orbit the centers of their galaxies in similar ways. For example, the entire solar system revolves around the center of the Milky Way galaxy at a velocity of about 250 kilometers per second. It completes one circuit of its orbit once every 250 million years.

Data obtained from the orbital motion of the sun allows astronomers to compute the mass that lies within the sun's orbit. Though matters are a bit more complicated than they are in the case of a planet that orbits a star, there are no real difficulties. In fact, the mass of the galaxy within any orbit can be found. For example, if a star is 30,000 light-years from the galactic center, observations of the star's motion allow astronomers to compute all of the mass contained in the galaxy out to that distance. In effect, the star orbits a "smeared-out" mass that includes the dense galactic core and all of the stars that are nearer to the core than it is.

Watching a galaxy to see how long it takes stars to complete their orbital circuits is obviously not a very practical endeavor. One would have to remain at the telescope for millions of years in order to carry out such a task. Fortunately, there is a simpler method of finding galactic mass. Since all of the stars of a galaxy that are located at the same distance from its center will revolve at the same velocity, one need only look at a galaxy edge on to determine how rapidly the stars are moving. The velocities of the stars can be calculated from their redshifts or blueshifts.

If such a galaxy is observed, it will have an appearance that is somewhat similar to that of a rotating phonograph record that is viewed from the edge. If the record (or galaxy) is moving in a clockwise direction, the left edge will move away from the observer while the right edge moves toward him. The only difference of significance is that a galaxy is not as rigid an object as a phonograph record. In a galaxy, different rates of rotation will be observed for stars that are different distances from its center.

Few galaxies can be observed edge on. But this creates no problems. If a galaxy appears tilted when viewed from the earth, it is not difficult to compute the correction factors that must be used. Nothing more than simple trigonometry is needed.

When such observations are carried out, it is found that a great deal of dark matter exists within and around galaxies. For example, observations of the orbital motion of our sun indicate that it feels a gravitational force that is twice as great as that which would be exerted by the visible matter in the inner regions of the Milky Way. It appears that if we consider only that part of the galaxy that lies within the sun's orbit, dark matter and visible matter exist in about equal quantities.

As one goes farther and farther out from a galactic center—the center either of our own galaxy or of others—even larger quantities of dark matter are observed. Radio astronomers have measured the redshifts of radio waves* emitted by clouds of cool hydrogen gas that lie far from galactic centers. They have found that the farther out one looks, the greater the discrepancy between the rotation velocity that is observed and that which would be expected if only visible matter were present. This implies that most of the dark matter lies outside of the bright parts of galaxies. This is why it has become customary to speak of galactic "haloes" even though some of the dark matter lies close to galactic cores.

What could this dark matter be? Two possibilities, dust and gas, have already been eliminated. There are others that cannot be so easily ruled out, but which are not thought to be especially likely. The haloes could be made up of stars that are too dim to be observed, of planet-sized objects, or even chunks of ice. Gas emits radiation and dust scatters it, but any macroscopic object that had a size anywhere between that of a grain of sand and a planet would do neither. Consequently it would be invisible to astronomical observation. Astronomers have commented, jokingly, that the dark matter could even be composed of copies of *The Astrophysical Journal*.

However, these possibilities have few advocates. Astronomers know of no processes that would cause rocks, ice chunks, dim stars, or starless planets to collect in regions so far from galactic centers. Even if this happened, it does not seem very plausible that the numbers of such objects could be so great that they would outweigh the visible matter in galaxies by a factor of 10 or more.

Furthermore, there are theoretical arguments which seem to indicate that the dark matter can be made up of none of these things, that it must be unlike any kind of matter that we know. The dark matter, these arguments say, is not even made up of atoms.

* Since "redshift" is defined to be a change in wavelength, the term can be applied to any kind of radiation; its use does not necessarily imply that any visible light is present.

When these arguments were first proposed, it was not the intention of their authors to say anything about the nature of the dark matter at all. On the contrary, they were attempting to find an indirect method of estimating the mass density of the universe in order to determine whether the universe was open or closed.

According to the big bang theory of the origin of the universe (to be discussed in detail in the next chapter), most of the helium that exists today was synthesized shortly after the universe was created. The big bang theory also implies that all of the *deuterium* (an isotope of hydrogen in which the nucleus contains one proton and one neutron rather than the single proton) that is observed was created at the same time.

Astronomical observations have established that the universe is about 75 percent hydrogen and 25 percent helium by weight. Heavier elements exist only in trace amounts.* Most of this helium must have been synthesized in the big bang. The universe has not existed long enough for more than a small fraction of the helium to have been created in the nuclear reactions in the interiors of stars.

Observations also indicate that the ratio of deuterium to ordinary hydrogen is about 20 to 30 parts per million. All of this deuterium must have been created in the big bang. The nuclear reactions that go on in stars do not create deuterium; they destroy it. Since the proton and neutron that make up the deuterium nucleus are weakly bound to one another, it does not take much energy to break the nucleus apart.

Measurements of the concentrations of helium and deuterium in the universe make it possible to calculate the rate at which the universe was expanding early during its history. When the expansion rate is compared with the matter density during that era, one obtains information that can be used to calculate what the matter density should be today.

The calculation is based on the idea that if the early universe had been expanding more slowly, or if it had been denser, more helium would have been produced. A greater concentration of matter would have caused the primordial hydrogen to be con-

---

* This seems to imply that we and the planet on which we live are nothing more than cosmic impurities.

verted into helium at a faster rate. On the other hand, a faster expansion, or a matter density that was less, would have caused a smaller amount of hydrogen to combine.

We needn't concern ourselves with the details of the calculation. The essential point is simply that the amount of helium in the universe is related to matter density. It is related to the matter density that existed 15 billion years ago, and it is related in a slightly more complicated way to the matter density that exists today. Knowing what the percentage of helium is should tell us about how much mass the universe contains.

There is also a relationship between the amount of deuterium that is present and the density of matter. In this case, matters are a bit more complicated. Not only can deuterium nuclei easily be broken apart into their constituent protons and neutrons, but they may also combine with one another to form helium nuclei. However, the principle is the same. The denser the early universe was, the more reactions would have taken place. Thus the deuterium-hydrogen ratio should also give us an estimate of the density of the universe 15 billion years ago, and an estimate of the density of the universe today.

When the calculations are carried out, one obtains the result that the matter that is present in the universe should have a density that is many times smaller than the critical value. This result is consistent with measurements of the amount of visible matter which is observed, and when these calculations were originally done, it was thought that this implied that the universe must be open.

However, as we have seen, no more than a tenth of the mass that is present in clusters can be directly observed. No one knows how much dark matter there is between clusters; however, it is possible to conclude that there is at least 10 times as much dark as visible matter in the universe. The minimum amount of dark matter that could be present is just barely consistent with the matter density predicted by the arguments based on the observed densities of helium and deuterium. In all probability, the universe contains much more mass than it theoretically should.

This suggests that we should ask whether the mass-density argument contains any loopholes. When we look at it closely, we find that it indeed does. The helium-deuterium argument really says nothing about the total amount of matter that should be present in the universe; it gives only the amount of *baryonic matter* that should be present. Baryonic matter is matter that is made up of baryons, such as neutrons and protons. It is ordinary matter, in other words. Of course, ordinary matter contains electrons too, and electrons are leptons, not baryons. But when one is talking about mass densities, the electrons can be neglected because they weigh so little.

Although observational and theoretical uncertainties make it difficult to be absolutely certain, it appears likely that the dark matter that makes up 90 percent or more of the mass of the universe is not ordinary, baryonic matter. If this conclusion is correct, the hypothesis that the dark matter is composed of wandering planets, ice chunks, rocks, or dim stars can be ruled out. Since there is no apparent reason why such objects should surround galaxies anyway, astronomers are not unhappy with this conclusion.

Then what is the dark matter made of? No one knows. The question of its nature is currently a topic of intense speculation, however, and a number of different possibilities have been suggested. A number of these will be considered in turn.

One of the first possibilities that was suggested, once it became apparent that the galactic haloes were not made of ordinary matter, was that they might be clouds of neutrinos. Some (but not all) of the grand unified theories suggest that the neutrino might not have zero mass, as had previously been believed. Around 1980, some experiments were performed which seemed to confirm this prediction.

In the spring of 1980, three University of California at Irvine physicists, Frederick Reines, Henry W. Sobel, and Elaine Pasierb, performed an experiment which seemed to indicate that one kind of neutrino could spontaneously transform itself into another. Their results suggested that an electron neutrino might sometimes turn into a tauon neutrino, and that the reverse process could also take place. Since theoretical calculations indicated that this kind of *neutrino oscillation* could take place only if neutrinos had mass, this implied that neutrinos might not be massless particles after all.

The experiment was one that was very difficult to perform,

and even the experimenters themselves did not believe that the results were conclusive. The experimental uncertainties were simply too great. In fact, when similar experiments were done elsewhere, the results were negative; neutrino oscillations were not seen.

In all likelihood, the Irvine physicists' result would have been quickly forgotten if, shortly afterwards, a group of Soviet physicists had not reported that they had detected neutrino mass. The Soviet experiment, which was carried out by a group of scientists at the Institute for Theoretical and Experimental Physics in Moscow, did not deal with neutrino oscillations. The Soviet physicists attempted to measure the neutrino mass directly. Reporting their results, they claimed to have found—with a certainty of 99 percent—that the mass of the electron neutrino was between 14 and 48 eV. The abbreviation "eV" stands for *electron-volt,* which is a small unit of energy. Since mass and energy are equivalent, physicists also use it as a unit of mass. The electron, for example, has a mass of 511,000 eV, while the proton weighs 938 MeV (938 million electron volts).

The Soviets made very accurate measurements of the energy of the electrons emitted in a certain radioactive decay process. They felt that if they could measure the energies of the electrons precisely enough, both the energy and mass of the neutrinos (which were not observed directly) could be inferred.

However, the neutrino mass that the Soviet scientists found was an extremely small quantity, less than a ten-thousandth of the mass of an electron. Other scientists expressed doubt about the results, pointing out that even very small experimental inaccuracies could produce spurious results. The Soviet experiment became a topic of controversy, and scientists found themselves unable to agree as to whether neutrino mass had been demonstrated or not.

In spite of all the uncertainties, the possibility of the existence of neutrino mass became a topic of discussion among scientists in a number of different fields. Some astrophysicists were especially intrigued by the idea. They began to wonder whether one might not be able to account for the dark matter by assuming that it consisted of neutrinos. To be sure, an individual neutrino did not weigh very much. However, the universe contained so many of them that it was indeed possible that neutrinos might account for the greater part of the universe's mass.

The existence of neutrino mass also seemed to suggest a possible solution to another outstanding problem. Astrophysicists had never been able to come up with a theory of galaxy formation that proved to be perfectly adequate. It seemed that whatever assumptions they made, their theories always produced predictions that were not quite consistent with the facts. Perhaps, the astrophysicists suggested, galaxies were basically clouds of neutrinos. If the neutrinos had clumped together early in the history of the universe, their mutual gravitational attraction would cause ordinary matter to collect around them. This ordinary matter could then condense and form galaxies inside the neutrino clouds.

For a while, the neutrino-mass theory seemed to be a promising one. It explained a number of phenomena that had previously mystified scientists. But in the end, the theory proved to be a chimera. Further experiments failed to confirm the existence of neutrino mass, and physicists found themselves becoming skeptical about its reality. The consensus was that the experiments which had seemed to show that neutrino mass existed were either inconclusive or wrong.

Meanwhile, theoretical calculations were carried out which seemed to indicate that even if massive neutrinos existed, they could not be used to explain the presence of dark galactic haloes. Scientists found that there were a number of reasons why the idea didn't really work. For example, calculations showed that massive clouds of neutrinos would give off X rays. These X rays were not observed. Next, it was found that the assumption that most of the mass of the universe resided in neutrinos led to a calculated expansion rate for the universe that was not consistent with observations. Finally, computer simulations indicated that neutrinos would not condense into clouds in the right way; they would form clumps that were unlike the galaxies that were actually observed.

If the dark haloes were not made of ordinary matter, and if they were not clouds of neutrinos, what were they? The astrophysicists who continue to study the problem aren't sure. However, they have concluded that the haloes are made up of *cold matter* of some kind. Cold matter is matter that is composed of particles that move through space at relatively low velocities. Neutrinos, on the other hand, are a kind of *hot matter*; they rush about in a more violent manner. The reason that the cold-matter hypothesis is now preferred is that since cold-matter particles have a different kind of velocity distribution, many of the problems that were associated with the neutrino hypothesis can be avoided.

Some physicists have suggested that the dark matter might be made up of photinos or gravitinos, two particles whose existence is predicted by supergravity theories. Alternatively, it might be composed of *magnetic monopoles* or *axions*, particles whose existence the grand unified theories predict. A magnetic monopole is an isolated north or south magnetic pole. If they exist, monopoles must be very heavy; they would weigh the equivalent of more than 10^{16} protons or neutrons, or about as much as a bacterium.

Axions are bosons that interact with ordinary matter in a very weak way (assuming, of course, that the predictions are correct, and that they really exist). They have a very small mass, about a hundred-thousandth of that of the electron. However, if the theories that predict their existence are correct, enough of them should exist to account for the dark matter. The gravitational effects of a cloud of axions would be similar to those produced by a dust cloud. However, axions would not scatter light and produce "cosmic smog" the way that dust particles would.

The only trouble with these cold-matter hypotheses is that none of these particles have been detected experimentally. Their existence is predicted by theory, but, as we have seen, the grand unified and supergravity theories must be considered to be very speculative in nature. Any theory which makes use of the assumption that one or more of these various kinds of particles is real could wind up meeting the same fate that the neutrino theory encountered.

If monopoles exist, they would be hard to detect because such massive particles could not exist in great numbers. If they did, the universe would have much more mass than is observed. At best, only about 200 monopoles strike a square kilometer of the earth's surface in a year. Attempts have been made to observe these particles, but the experiments have so far not been successful.

On the other hand, axions would be hard to detect because their mass is so small, and because their interactions with ordinary matter would be correspondingly feeble. A number of experiments have been performed in which attempts were made to look for axions, but these have not been successful either. Finally, as far as I am aware, no experiments designed to detect photinos or gravitinos have even been proposed. At the moment, the outlook for verifying their existence does not seem to be terribly bright.

It appears that at the moment, the best that physicists can do is to suggest that the dark matter is made of particles whose existence has not been verified. Thus one must conclude that little progress is likely to be made toward finding out what the dark matter is until some significant theoretical or experimental advances are made. At present, all that can really be said is that at least 90 percent of the universe is made up of a kind of matter that we cannot see, whose makeup we do not know.

There is yet another possibility. Since it is the most speculative of all, I have left it for last. The dark haloes could conceivably be made of shadow matter. Since shadow matter interacts with ordinary matter only through the gravitational interaction, it could not be seen. Furthermore, there is no reason why it could not exist in greater quantities than ordinary matter. Although the possibility can't be considered to be especially likely, at least not until we have better reasons for thinking that there really is such a thing as shadow matter, it does seem to be possible that we live in a shadow matter universe which contains relatively small amounts of matter of the "ordinary" visible variety.

6

Being and Nothingness

As we have seen, the universe is not what it appears to be. For thousands of years, human beings have been looking up into the night sky and have seen a cosmos that was bright and static. But the universe of modern physics bears little resemblance to the one that human beings have thought they have seen. The universe is not static; it is in a state of rapid expansion. Most of it is not bright; the bulk of the matter that it contains is dark, invisible, and of unknown composition.

But these are not the only surprising discoveries that have been made about the universe. One discovery has followed another, with the result that scientists' conception of the nature of the cosmos has been changed again and again, with the result that the universe as it is conceived today bears little resemblance to that which scientists thought they saw at the beginning of the twentieth century. The modern conception of the universe is not even very much like the one that Einstein intuited when he tried to understand its nature by applying the ideas of general relativity.

For example, the universe, which can be naively defined as "everything that exists," comes very close to being nothing. Furthermore, it is possible that this "everything-nothing" universe of ours is only one of an infinite number. Today, many scientists believe that the universe popped into existence as a result of a kind of quantum fluctuation. If it did, there is no reason why the same process cannot have taken place numerous times. The universe that we see may not be "everything that exists" after all. There may be countless others like it.

When one looks at the mass and energy content of the universe, a very strange thing is observed. When the universe's mass and energy densities are added together (since mass and energy are equivalent, there is nothing to prevent us from doing this), one obtains an odd result. The gravitational energy in the universe is negative, and there seems to be just about enough of it to balance out the positive contribution of the mass.

The idea of negative energy isn't a particularly esoteric one. For example, a hydrogen atom, which consists of an electron and a proton, weighs a little less than the electron and proton do when they are not bound in an atom. Similarly, an atomic nucleus has a mass which is a little less than the combined masses of its constituents. In each case, the difference is equal to the amount of negative energy that the atom or nucleus contains.

In order to see why this energy should be negative, consider an electron in an atom. Since there is an attractive force between the negatively charged electron and the positively charged nucleus, energy must be expended if the electron is to be moved out of the atom. Thus a free electron should have a greater amount of energy than one that is circling a nucleus.

Suppose that the electron and the nucleus are so far away from one another that the attractive force they experience is negligible. In this case, we can say that the energy of the pair is zero. But if the electron-nucleus pair has less energy than this when the electron is bound in an atom, we must conclude that the energy is negative in this case. This result, incidentally, is confirmed by experiment. The change in mass caused by the presence of this negative energy has been measured.

The situation with respect to gravity is very similar. Two bodies will have zero gravitational energy when they are very distant from one another. Since it takes energy to move two gravitating masses apart, the gravitational energy of the bodies must be less than zero whenever they are close enough to one another to feel an attraction. When numerous gravitating bodies are bound to one another, the amount of negative energy is much greater.

When one adds up all the mass in the observable portion of the universe, a very large number is obtained. When one calculates the amount of negative gravitational energy that exists in the same region of space, another large number is obtained. As far as we can tell, the two numbers are about equal.

Although this result is intriguing, it is really not possible to draw any conclusions from it. No one knows whether the sum of these two numbers is positive, negative or zero; the uncertainties are too great. Even if the sum of the mass and energy were exactly zero, it would not be entirely clear what the implications of this were. The role that energy plays in the general theory of relativity is a strange one. It appears that energy is not conserved in an expanding universe. Furthermore, grave mathematical difficulties are encountered if one attempts to define what the energy content of the universe is. In a closed universe, this cannot be done. The words "total energy" are meaningless; the concept simply doesn't apply to the universe as a whole. In an open universe, matters are just as bad. Since an open universe is spatially infinite, its total energy would have to be infinite too.

Adding matter and energy together to get zero is thus a mathematically dubious procedure. However, the nearness of the matter-energy to zero is a bit unexpected. Offhand, one would think that something as imposing as the universe should contain a quantity of matter-energy that was very great.

This is not the only surprise that anyone who looks at contemporary theories of the universe is likely to encounter. Some of the results that have been obtained are astonishing indeed. For example, there is the *inflationary universe* theory, which seems to suggest that the universe, and everything in it, may have come into existence from nothing.

The inflationary universe theory implies that, early in its history, the universe went through a brief period of rapid, *inflationary* expansion. According to the theory, most or all of the matter and energy in the universe were created during this period of inflation. Originally, the universe was nothing more than a tiny bubble of space-time that may have contained something like twenty pounds of matter, or perhaps no matter at all.

The inflationary universe theory is an elaboration of the big bang theory of the origin of the universe. So perhaps it would be best to begin by discussing the development of the latter. This will not only make it somewhat easier to understand what the inflationary theory is all about, but it will also allow me to show how dramatically scientific conceptions of the nature of the universe have changed during the twentieth century.

The term "big bang," which was first used by the British astronomer Fred Hoyle on a radio broadcast in 1950, is a reference to the fact that the universe is thought to have exploded into existence approximately 15 billion years ago. According to the theory, the expansion of the universe implies that all of the matter that it contains must have been compressed into a very small volume at some point far in the past. It is inferred that at one time, everything was packed together in a very dense, very hot state. As we shall see, there are various kinds of observational evidence which confirm that this inference is correct.

The first scientist to suggest that the observation that the universe was expanding could be used to draw conclusions about its origin was the abbé Georges Lemaître, a professor of relativity and the history of science at the University of Louvain. In 1933, Lemaître theorized that the universe originally consisted of a "primeval atom," in which all the matter that makes up the present-day universe was concentrated. At some point in time, according to Lemaître's theory, this primeval atom disintegrated, sending matter flying off in every direction.

The idea was an intriguing one, but it really didn't explain very much. A scientific theory is considered to be useful only if it can account for a wide variety of phenomena. In particular, a theory of the origin of the universe should not only explain the expansion, but it should also tell us why the universe has the chemical composition that is observed. It should explain why there is about 75 percent hydrogen and 25 percent helium, and why heavier elements should exist only in trace quantities.

Significant theories always have a wide-ranging character; it is this that allows us to distinguish them from ad hoc explanations.

This character was precisely what Lemaître's primeval atom theory lacked. About the only significant prediction that could be derived from it was that the universe was expanding, which was something that was already known.

Nevertheless, individuals who are capable of having significant scientific insights are rare, and Lemaître's work must be considered to be of great importance. After all, Lemaître was the first to point out just what the implications of the expansion of the universe were. This is sufficient to allow him to be considered the father of the big bang theory.

Some fifteen years passed before the primeval atom theory was improved upon. Then, in 1948, a paper appeared in the American scientific journal *Physical Review*. The names of three authors, Alpher, Bethe, and Gamow, appeared under the title. In reality, the paper had been written by a Soviet émigré scientist, George Gamow, together with his student, Ralph Alpher. Cornell University physicist Hans Bethe didn't even know about the paper until he saw it in print. Gamow had added Bethe's name as a kind of practical joke, giving the paper an authorship that was a pun on the first three letters of the Greek alphabet.

According to the Alpha-Beta-Gamma theory, as it came to be known, the universe did not begin as a cool primeval atom, but as an intensely hot fireball. Initially, according to Alpher and Gamow, there existed temperatures in excess of a billion degrees Celsius. The energy created by the great heat caused the particles which populated the universe to participate in various kinds of nuclear reactions. It was in these reactions, the authors of the theory said, that many of the chemical elements that exist today were created.

The idea that the universe should once have been very hot is quite a reasonable one. All substances heat up when they are compressed, and cool when they expand. This is the reason, by the way, that aerosol sprays feel cool. The gases in an aerosol can expand and cool when they are released. The expansion of the universe is an analogous process. We have every reason to think that the universe has been cooling off for the last 15 billion years.

Alpher and Gamow's intention, in proposing the big bang theory, was not only to explain the expansion of the universe, but also to show how *nucleosynthesis,* or the formation of the elements, could have taken place. According to the theory, the universe was originally composed of neutrons. These neutrons decayed into protons, electrons, and neutrinos by the beta decay process. The high temperatures that existed at this stage would then have provided the energy needed to cause the neutrons and protons to combine to form atomic nuclei. Thousands of years later, these nuclei captured free electrons to form the atoms that make up ordinary matter today.

And then, what appeared to be a flaw was discovered in the theory. Detailed calculations showed that although helium nuclei could easily have been created in the big bang fireball, the nuclei of heavier elements would not have been formed. There seemed to be a kind of nuclear "energy gap" that not even the high temperatures of the big bang could have overcome. Thus the big bang theory seemed to imply that the universe should contain no carbon, nitrogen, oxygen, copper, iron, or any of the other elements with which we are so familiar. For that matter, it seemed that the theory implied that there should be no Ralph Alpher or George Gamow. After all, one cannot construct a human being from hydrogen and helium.

A solution to the problem was soon found by Fred Hoyle, who theorized that the heavy elements could have been created in the nuclear reactions that take place in the interiors of certain stars. According to Hoyle (no pun intended), if some of these stars were later torn apart by supernova explosions, these elements could be scattered through space and incorporated in second-generation stars, and in the planets that formed around them.

Today, Hoyle's theory is universally accepted. There is still much that scientists do not understand about supernovae, and no one knows precisely how large a star must be to explode. However, it has been established that many massive stars explode in this manner when their nuclear fuel is exhausted, and supernova remnants, such as the well-known Crab nebula, have been observed and studied.

At first, the big bang theory attracted little attention, even after Hoyle had cleared up the problems about the formation of the elements. Apparently, most scientists did not really believe that it was possible to deduce anything about events which had presumably taken place billions of years in the past. But then, in

1964, they suddenly realized that the theory did have to be taken seriously. In that year, two American scientists, Arno Penzias and Robert Wilson, observed the light that had been emitted in the big bang fireball.

It wasn't light that they saw, however, but rather radio waves. The expansion of the universe causes any radiation that travels over large distances to be redshifted. The longer it has been traveling, the greater the redshift will be. One should expect that the shift would be great indeed in the case of radiation that has been traveling through space for 15 billion years.

The radio waves that Penzias and Wilson detected fall on the earth from all directions of space. This is exactly what one would expect. After all, the big bang did not take place at any particular point in space. It happened everywhere. The universe may be infinite, or it may be finite. In either case, the big bang filled all of existing space. Then, as space expanded, matter flew outward with it.

This point is worth emphasizing. Many of us have a natural inclination to visualize the big bang by comparing it to terrestrial explosions. However, viewing it in this manner is an error. There was nothing outside the big bang: no space, no time, no energy, no particles of matter. The earth itself is located inside the expanding space that was created in the initial explosion.

Since Penzias and Wilson discovered the radio waves from space, or *cosmic microwave background radiation*,* astronomers and physicists have been elaborating upon the big bang theory, and seeking additional ways to test it. They have been remarkably successful. The theory is now so well established that it is difficult to entertain doubts about its validity, and physicists now believe that they can speak confidently of events that took place less than a billionth of a second after the universe began. Some of them think that it is possible to push back even further. Some current scientific speculation deals with events that might have happened only 10^{-43} seconds after the creation of the universe.

In discussing the big bang theory, I have been using such expressions as "creation," "beginning," and "origin of the uni-

---

* It is called "cosmic" because it comes from space, "microwave" because it is observed at microwave frequencies, and "background" because it is everywhere.

verse." However, these terms should not be interpreted literally. Scientists do not know what happened before the big bang, or even if there was such a thing as "before." There has been some speculation to the effect that the universe came into existence as a kind of quantum fluctuation (I will discuss this idea later in the chapter), but no one knows for sure. It is also conceivable that the universe existed in some manner or another before the big bang took place. For example, the universe may, at one time, have been in a state of contraction. It might have become more and more compressed until it reached a point where the compression could continue no longer. If this happened, it could have bounced back into the state of expansion that we observe today.

If, on the other hand, the universe was created in the big bang, it may be meaningless to speak of what happened before this event. Space and time themselves may have been created in the big bang. The cosmos may have no previous history.

The reason that science is unable to describe events which might have taken place before the big bang is that it is only possible to go so far back in time before all the known theories of physics break down. The only way to understand the nature of the universe is to use the general theory of relativity. But general relativity is no longer valid when matter densities are very great. There is a point where quantum effects become so important that the theory would have to be modified. Unfortunately, scientists do not know what these quantum effects would be like, or how they would influence gravitational interactions.

It is believed that general relativity should remain valid down to a time of 10^{-43} seconds after the beginning of the expansion. The figure of 10^{-43} seconds, incidentally, is one that we have encountered before; it is the Planck time. To be sure, theories about events which presumably took place after the Planck time are often somewhat speculative. However, it is at least possible to speculate about events during this era. When one considers the era before the Planck time, it is hardly possible to do even that.

It is thought that space and time themselves were subject to quantum fluctuations during the Planck era. Until a way is found to understand what these fluctuations were like, and to modify the general theory of relativity accordingly, little or no progress can be made. The problem of dealing with time may prove to be an especially difficult one. The theories of physics, including quantum mechanics, generally describe processes which take place in time. Time is regarded as something that is smooth and continuous. But this smooth, continuous time would have to be replaced by something else in a theory of quantum gravity, and it is not obvious what this "something else" would be. After all, one cannot say that fluctuations of time take place in time. The circularity of such a statement makes it meaningless.

There exist a number of experimental confirmations of the big bang theory. The evenness of the cosmic microwave background is one. The background radiation varies in intensity by less than one part in ten thousand when it is measured at different points in the sky. If the radiation was produced by processes that had nothing to do with the big bang, one would expect that it would come from certain specific directions, or be more intense in one area of the sky than another. The fact that it comes from everywhere confirms the hypothesis that it was produced by events that happened everywhere, in other words, in the big bang.

The background radiation also varies in intensity with wavelength in precisely the manner that theory predicts. The radiation seems to be identical to that which would be produced by a blackbody with a temperature of 2.7 K (2.7 Celsius degrees above absolute zero; absolute zero is the lowest possible temperature, the one at which all molecular motion stops).

Measurements of the concentrations of helium and deuterium that are present in the universe also confirm the big bang theory. As I pointed out in Chapter 5, the amount of helium in the universe is much greater than the quantity that could have been produced in stars since the beginning of the universe. Deuterium cannot be made in stars at all; the high temperatures that exist in stellar interiors cause deuterium to break apart.

There are also theoretical reasons for believing that the universe began with a big bang. If one assumes that general relativity is a correct theory, and if one makes certain reasonable mathematical assumptions, it is possible to prove theorems which indicate that all the matter and energy in the universe should originally have been compressed into a *singularity* of zero volume. In other words, the density of matter and energy were originally infinite.

One does not have to take the idea of an infinite compression of matter literally. These theorems are based on general relativity, and it is known that general relativity breaks down at the Planck time. However it is possible to project back at least this far, and to conclude that at 10^{-43} seconds after the beginning, the universe must have been in a highly compressed state, from which it has been expanding ever since.

Although the big bang theory has had a number of outstanding successes, there are also a number of difficulties associated with it. For example, the theory does not explain why the average curvature of space should be so close to zero. Astronomers have not been able to determine whether we live in a closed universe of positive spatial curvature or in an open universe of negatively curved space. They can only tell that the universe's mass density is somewhere between a tenth of the critical value and 10 times that amount.

Admittedly, a factor of 100 ($\frac{1}{10}$ is a hundredth of 10) does not seem so very small. For that matter, one might think that a universe that had a mass density that was a tenth of the critical amount would not exactly be teetering on the borderline. But as a matter of fact it would be. In fact, the closeness of the mass density to the critical one is astonishing.

This point can be best understood if we look not at the mass density, but at the expansion rate. After all, the two quantities are related. One can say that an open universe is one that contains too little gravitational mass to halt the expansion. Or one can say that an open universe is one that expands so rapidly that gravitational retardation will never bring the expansion to a halt. The two statements are perfectly equivalent.

If the universe has between $\frac{1}{10}$ and 10 times the critical mass density, then it is possible to calculate that the universe must have been expanding at just the right rate, to an accuracy of one part in 10^{60}, at the Planck time. If the expansion had been just a little less rapid, gravity would have quickly halted the process and have caused the universe to collapse again in what has been called a "big burp." If the expansion had been a tiny bit faster, matter would have dispersed so rapidly that galaxies could never have been able to form. The relationship between density and expansion rate indicates that the fine-tuning must have been precise indeed to create the kind of universe that is observed today. The big bang theory does not explain how this fine-tuning came about.

Another problem with the theory is that it fails to explain why the universe should look pretty much the same in every direction. To be sure, this sameness, or *isotropy*, is not very obvious at first. Galaxies are not uniformly distributed; they are grouped into clusters and into superclusters. In recent years, astronomers have discovered that the universe also contains huge "holes in space" that are almost devoid of galaxies.

However, when one ignores the irregularities that are observed in our galactic neighborhood and looks at the universe as it appears at distances of billions of light-years, the distribution becomes remarkably uniform. A good analogy is provided by the distribution of sand grains on a beach. To a creature the size of an ant, a beach appears quite irregular; individual sand grains take on the appearance of boulders. But to a human being who looks out over a relatively long distance, a beach appears to be a uniform expanse of sand.

The isotropy of the universe becomes even more striking when one looks at the microwave radiation background. The intensity of the background radiation does not vary by more than one part in ten thousand. This fact is astonishing when one considers that the background radiation that comes from opposite sides of the sky was emitted some 15 billion years ago by portions of the universe that were apparently never in causal contact.

In order to understand the significance of this, it is necessary to know something about horizons in the universe. For example, no matter how powerful our telescopes become, it will never be possible to see more than about 15 billion light-years into space. The reason that it will not is that there has not been enough time since the beginning of the universe for light to travel farther than that. It is likely that there are stars and galaxies situated at greater distances. However, until billions of years pass, it will not be possible to observe them.

For that matter, nothing beyond the cosmic horizon can possibly affect us in any way. According to the special theory of relativity, no signal or causal influence can travel at velocities greater than that of light. The galaxies beyond the horizon cannot exert any influence upon us now, and they cannot have done so at any time in the past. After all, in the past, the horizon distance was even shorter.

The microwave radiation was emitted when the universe was about 500,000 years old. At that time, portions of the universe more than 500,000 light-years away from one another had presumably never been in causal contact. Yet, when we look at opposite sides of the sky, we see microwave radiation coming from regions that were much farther away from one another than that. If these two regions of space can never have known of one another's existence, how can they mirror one another so exactly? The most obvious explanation is that they were both affected by some kind of "smoothing out" process that made radiation emission uniform. But we know from special relativity that this is impossible. The "smoothing out" influences would have had to travel much too fast. Finding that the microwave radiation is the same all over the sky is like filling the opposite ends of a bathtub with hot and cold water, and then discovering that the temperature has become perfectly uniform after only a fraction of a second has passed.

There are yet other phenomena that the big bang theory leaves unexplained. For example, it does not tell us why stars should congregate together into galaxies. There is no problem with star formation itself. Calculations show that the mutual gravitational attraction of the gas atoms that fill interstellar space should cause the gas to contract. As it does so, temperatures rise, and the gas begins to glow. As the contraction continues, temperatures and pressures increase still further until the nuclear reactions that take place in the interiors of stars begin.

Gravitational attraction presumably causes stars to congregate into galaxies also. But when the relevant calculations are carried out in detail, problems appear. It appears that stars would indeed collect into groups. However, the stars' mutual gravitational attraction would not be sufficient to bring about the creation of galaxies, which are huge structures 10,000 to 100,000 light-years across, and which contain as many as a trillion stars. Even the presence of interstellar gas or cold matter does not solve the problem. According to the big bang theory, galaxies should not exist.

If galaxies were to be created there must have been regions of the universe where the density of matter was a little higher than average. If random fluctuations created regions where the density was only about a hundredth of a percent higher than normal, gravity would have caused additional matter to collect in these regions. Then, when star formation began, greater-than-average numbers of stars would be created where density was high.

It is hard to imagine that galaxies could have been created in any other way. Yet, according to the big bang theory, this scenario cannot be correct. The expansion of the universe would have caused the condensations of matter to dissipate. According to the theory, either galaxies should not have formed at all, or the time required for their creation should be much greater than the present age of the universe. The only way to avoid this result is to assume that the big bang contained "lumps" of some kind that evolved into galaxies later. However, no one knows of any process that could have created such lumps. To invoke their existence is to invent an ad hoc hypothesis to patch up a theory that does not quite work.

The three problems associated with the big bang theory have names. The first is called the *flatness problem*. If the density of matter in the universe is close to the critical value, this implies that the average curvature of space must be close to zero. Zero-curvature space is called *flat*. The second is called the *horizon problem*. The isotropy of the universe implies that regions that are beyond one another's cosmic horizons are more alike than they should be. The third is called the *galaxy formation problem*, for obvious reasons.

Before I discuss a possible solution to these problems, it might not be a bad idea to consider another puzzle, that of the nonexistence of *antimatter*. Antimatter is matter that is made of antiparticles. For example, an atom of antihydrogen would be composed of a positron and an antiproton. Its properties are identical to those of an atom of ordinary hydrogen. It emits and absorbs radiation at exactly the same wavelengths, and it has the same mass. With one exception, there is no way that matter and antimatter could be distinguished.

The exception is quite an important one, however. If matter and antimatter came into contact, they would annihilate one another, releasing large quantities of energy in the process. The energy created in a matter-antimatter explosion would be many times greater than that produced by a hydrogen bomb; in a nuclear explosion, only a small fraction of the matter that is present is converted into energy.

The fact that matter-antimatter explosions are not observed tells us that there is no antimatter in the solar system. If there were, antimatter meteorites would release large amounts of energy when they struck planets, and radiation would be observed when particles of ordinary interplanetary gas came into contact with particles of antimatter.

It is also apparent that our galaxy can, at most, contain only minute quantities of antimatter. If antimatter were present to any appreciable extent, the collisions between interstellar dust, gas, and stars would release large quantities of gamma rays, which would certainly be detected. It is virtually impossible to imagine any process that could keep matter and antimatter segregated.

Conceivably, entire galaxies could be made of antimatter. However, if such galaxies exist, they have not yet given us any sign of their presence. Galaxies collide from time to time. If a matter galaxy and an antimatter galaxy came into contact with one another, large quantities of radiation would again be released. Since this has not been observed, it seems reasonable to conclude that the universe is made entirely of matter, at least until we have some evidence to the contrary.

The prevalence of matter and the apparent nonexistence of antimatter is something that requires explanation. The particles of matter that exist today could very well have been created out of pure energy shortly after the beginning of the big bang explosion. But if they were, one would think that matter and antimatter would have been created in equal quantities. After all, this is what happens under terrestrial conditions. When particles and antiparticles are created in high-energy accelerators, their numbers are always equal.

One can always assume that the universe must have contained more matter than antimatter from the very beginning. But this is

hardly a very satisfying solution to the problem. Making this assumption is like saying there is more matter than antimatter because there has always been more matter than antimatter.

The fact that matter and antimatter can be created in the
laboratory suggests that perhaps the quantum field theories that
describe the behavior of subatomic particles might provide a solution to the problem. When one looks at these theories in detail, it
is found that neither the electroweak theory nor QCD is of any
help. According to these theories, the creation of matter and
antimatter should be a symmetrical process. There should be just
as much of one as there is of the other.

Since the two theories that constitute the standard model suggest no answers, it is necessary to look at the predictions made by
the more speculative grand unified theories. There is really no
reason not to do this. Perhaps these theories have not been experimentally confirmed. However, this should not stop us from trying to see what their implications are. In fact, if we find that the
GUTs do suggest a reason why only matter should exist, this
would be evidence that one or another of these theories may be
correct.

As I noted in a previous chapter, the GUTs seem to imply that
the proton is unstable. Even though this prediction has not been
confirmed, it suggests that perhaps protons can be created as well
as destroyed. In particular, perhaps there is a way to create protons without creating antiprotons too. If this can happen, then
perhaps there were events taking place in the early universe that
could have been responsible for the preferential creation of matter over antimatter.

According to the grand unified theories, such events can indeed take place. The GUTs predict that there should exist a class
of twelve particles that is designated by the letter X. The X particles are bosons that have the property that they can turn quarks
into leptons, or leptons into quarks. The particles have a mass of
approximately 10^{15} GeV. A GeV is a billion electron volts. The
reason that the symbol "BeV" is not used is that the word "billion"
does not have the same meaning in the United States that it has in
Great Britain and in most European countries. Our "billion" is a
thousand million. The European "billion" is a million million. On
the other hand, there is universal agreement that the symbol "G"

(for "giga") can be used to represent the number 10^9 (our "billion").

The quantity 10^{15} GeV is a huge mass. The highest energies—and thus the greatest masses—that can be produced in particle accelerators are only about 100 GeV. As a result, there is little chance that X particles will ever be created and detected by experimental physicists. However, scientists may be able to obtain indirect evidence of their existence if they are ever able to observe proton decay. According to the grand unified theories, a proton will decay if two of its constituent quarks exchange an X particle. The quark that emits the X will be transformed into a lepton, while the quark that absorbs the X will change into an antiquark. The three quarks that originally made up the proton are thus changed into a lepton and a quark-antiquark pair, in other words, into a lepton and a meson (a positron and a pion, for example).

The mass of a proton is a little less than a GeV. One might wonder how a proton could possibly contain an X particle, which weighs 10^{15} times more than it does. But there is really no difficulty. Heisenberg's uncertainty principle allows high-mass particles to exist for short periods of time anywhere. However, the more massive the particle, the shorter the time that it can exist. In the case of a virtual X particle, the time is about 10^{-39} seconds. It is obvious that an X particle that is emitted by a quark will not have much time to seek out a second quark. In most cases, it will be absorbed by the particle that emitted it before it has a chance to find and interact with another. This explains why proton decay should be such a rare event.

The exchange of an X can cause a proton (or a neutron, since neutrons are made of quarks too) to decay. The process does not cause protons to be created. At first glance, the GUTs seem to have little to say about the imbalance between matter and antimatter that is observed in the universe. But this is really not the case. There is every reason to think that shortly after the creation of the universe, sufficient energy was available to create real X particles. If there was, events could have taken place that are no longer possible today.

Since the universe has grown progressively cooler as it has expanded, there must have been a time when it was hot indeed. In particular, one can calculate that when the universe was 10^{-32}

seconds old, it had a temperature of about 10^{28} K. But temperature is nothing more than a measure of the average energy of the particles that make up a substance. For example, the molecules that make up hot water move about more rapidly and energetically than the molecules in water that is cold. One can calculate that at a temperature of 10^{28} degrees there should have been enough energy available to cause X particles to be created in great numbers.

At this point, the skeptic is likely to ask whether one can really speak meaningfully of events which supposedly took place 10^{-32} seconds after the universe began. He is likely to wonder if there is any way to verify that a temperature of 10^{28} K ever existed, and whether it is really possible to explain the behavior of particles such as the X, which will probably never be observed.

A skeptic who made these objections would have a point. However, the only way to find out whether such skepticism is misplaced is to forge ahead and to try to determine what the GUTs have to say about processes which might have taken place in the early universe. If we find that consideration of these processes allows us to make theoretical predictions about the nature of the present-day universe, then these predictions can be checked against observations. If there is a reasonably good fit between observation and theory, then this constitutes evidence that our theorizing was not so unrealistic after all. If we find that our analysis of events which took place at a time of 10^{-32} seconds works, this implies that perhaps we really can say something about conditions in the early universe.

The GUTs imply that the existence of real X particles in the early universe can indeed lead to an excess of matter over antimatter. The reason for this is that when X particles decay, they do not decay into equal amounts of matter and antimatter. For example, X's may decay to two-thirds matter and one-third antimatter. On the other hand, anti-X's (or $\overline{X}$'s; a bar is often used to designate an antiparticle) may decay to two-thirds antimatter and one-third matter. Thus if the X's and $\overline{X}$'s decayed at the same rate, equal amounts of matter and antimatter would again be produced. However, according to the GUTs, these decay rates do not have to be the same. If they are not, it is possible for more matter than antimatter to have been created in X and anti-X decays.

If this was the case, the kind of universe we see today—a matter universe—would have been created. For example, if a billion and one parts of matter are created for every billion parts of antimatter, then a bit of matter will be left over after the antimatter and a billion of the parts of matter have undergone mutual annihilation. If this theoretical description is correct, then it is this leftover matter that makes up the stars and galaxies that we observe today.

It cannot be said that this is a confirmation of one or another of the grand unified theories. The GUTs must still be considered to be speculative in nature, and the description of the creation of matter at least as speculative. However, the success of the GUTs in providing a hypothetical explanation for the preponderance of matter should give us some small bit of confidence in the theories. At the very least, we should be prodded into going on and trying to see what else the GUTs might tell us about the early history of the universe.

If one does this, the next thing that happens is that what appears to be a serious problem is encountered. The GUTs imply that large quantities of monopoles should also have been created in the early universe. When the numbers of these monopoles are calculated, one finds that the universe should contain enough monopoles to give it a mass density that exceeds the critical density by a factor of more than 300 billion. Not only does this imply that monopoles should outweigh ordinary matter by about 300 billion to one, it also implies that the universe should have ceased to exist long ago. A universe with so much mass would collapse upon itself in about 30,000 years.

In 1980, physicist Alan H. Guth of the Massachusetts Institute of Technology proposed his inflationary universe theory. One of his motivations for suggesting the theory was to see if there was some way that the production of so many monopoles could be avoided. When the theory was worked out in detail, it was found that not only did it successfully accomplish this task, but it also solved some of the problems associated with the standard big bang theory of the origin of the universe. The inflationary universe theory gave possible solutions to the horizon and flatness problems. Though it did not solve the problem of galaxy formation, it created hope that progress might be made in that direction.

Guth's theory was based on a consideration of events that might have taken place when the universe was 10^{-35} seconds old Since 10^{-35} is one-thousandth of 10^{-32}, we are speaking of a period somewhat earlier than the epoch during which the X particles decayed. According to Guth's theory, the universe underwent a period of rapid, inflationary expansion during the time that it was between 10^{-35} and 10^{-32} seconds old. During this time, according to Guth, the universe increased in size by a factor of 10^{25} or more.

In order to see why such an inflationary expansion should have taken place, it is necessary to know a few facts about the behavior of quantum fields and about the nature of gravity according to the general theory of relativity. I will examine the latter topic first.

In general relativity, gravitational forces* have a somewhat more complicated character than they do in Newton's theory. For one thing, gravitational masses are not the only things that create forces of attraction; the gravitational fields themselves do this also. For example, the mass of the sun causes gravitational forces to be exerted on the earth and the other planets; the gravitational fields that surround the sun add a bit to this attractive force. In general relativity, gravity itself gravitates.

Pressure also adds to the intensity of gravitational attraction. For example, the outer layers of the sun are held up by the pressure exerted by the gases that are heated by the nuclear reactions in the sun's interior. If these reactions were to cease, the sun would collapse like a balloon from which the air was gradually released. The existence of this pressure also adds a bit to the sun's gravity.

The fact that pressure should play such a role in general relativity seems to imply that a negative pressure would give rise to a repulsive "antigravity" force. In fact, the existence of negative pressure would produce a force like the one that was represented by Einstein's cosmological constant.

As we have seen, Einstein eventually discarded the idea of a

* It is true that, in the theory, the concept of "force" is replaced by that of curvature of space. But, as long as this is understood, one can revert to the former usage when doing so makes matters simpler.

cosmological constant. Modern scientists tend to think that this was a correct decision, at least as far as the present-day universe is concerned. If there is a cosmological constant today, it is so small that its effects are negligible. However, this does not necessarily imply that the constant was zero in the early universe. In particular, if there was negative pressure, the repulsive force associated with the constant would have been very real.

If there were any ordinary substances that exerted negative pressure, their behavior would be extraordinary indeed. If such a substance were placed in a flexible container, its presence would cause the walls of the container to bulge inward. If such a substance were mixed with the air in an inflated balloon, the negative pressure would counterbalance the positive pressure of the air and cause the balloon to become partially deflated.

No such substances are known. However, this does not necessarily imply that negative pressures cannot exist. In fact, they may be very real; under certain circumstances, such pressures may be created by quantum fields. They ought to be produced by the fields that exist in a vacuum.

Modern physics tells us that there is no such thing as "empty" space. Virtual particles and antiparticles are constantly being created and destroyed. They come into existence for short periods of time. Then they annihilate one another. The existence of such quantum fluctuations implies that the vacuum must have an energy. Furthermore, under certain conditions, it may be possible for the vacuum to exist in more than one energy state.

When Guth applied the grand unified theories to events which presumably took place 10^{-35} seconds after the creation of the universe, he found that the universe should have gotten temporarily stuck in what is called the *false vacuum* state. This terminology should not be taken too literally. The false vacuum is nothing more than an energy state of "empty" space that is not the lowest one possible.

When the universe was in the false vacuum state, the quantum fields associated with the vacuum exerted negative pressure. Since, according to general relativity, pressure creates gravitational force in an inward direction, negative pressure would have caused the universe to expand outward at a rapid rate. As the repulsive forces caused the universe to expand, matter and energy rushed in to fill the rapidly expanding space. Then, when the universe was about 10^{-32} seconds old, it underwent a transition to the lower, *true vacuum* energy state. The negative pressure disappeared, and the universe continued its expansion at a more leisurely pace.

The transition to the true vacuum state has been compared to *phase transitions,* such as the boiling or freezing of water. There is some similarity. When water freezes, for example, it makes a transition from one energy state to another. There is no change in temperature. When water freezes into ice, it will maintain a temperature of 0° Celsius until all of the water has been solidified. Only after it has been converted into ice will the temperature drop further. And yet, the properties of the water that is freezing change dramatically. The water molecules, which previously had moved about freely, align themselves in rigid ice crystals. At the same time, the volume increases; if it did not, ice would not float.

The idea that the properties of the entire universe should be able to suddenly change in a similar manner is a bit surprising. However, the idea should not be that difficult to understand; in fact, the transition to the true vacuum can be viewed as a kind of "freezing."

The idea that matter and energy should have streamed into existence as a result of the expansion is surprising too. However, if the GUTs are correct in their broad outlines, it must have happened. Negative pressure states have the curious property that the energy density of a given volume of space is constant. As a result, as the volume of space expanded during the inflationary period, more and more energy and matter would have been created. According to the inflationary universe theory, most—or perhaps all—of the matter and energy in the universe came into existence during this period of rapid expansion.

The inflationary universe theory implies that we must take the idea that the universe may have been created from nothing quite seriously. Before the inflationary expansion began, it may not have contained any matter at all. The universe may have begun as a bubble of empty, expanding space that came into existence by chance as the result of some Planck era quantum fluctuation. If it did, it could have continued to expand until it got caught up in the inflationary expansion that took place at a time of 10^{-35} seconds. After the inflation ended, it possessed huge quantities of matter and energy, and gradually evolved into the universe that we see today.

There is little that can be said about the creation of a space-time bubble, since we have no idea what the Planck era fluctuations were like. However, there is good reason to believe that a universe that originally contained no matter could expand. The equations of general relativity possess solutions that correspond to precisely such a universe. According to the theory, there can be such a thing as expanding empty space.

Some of the implications of the inflationary universe theory seem bizarre. However, as we have seen, many of the theories upon which contemporary physics is based seemed "crazy" at first. Thus we cannot judge the inflationary universe theory according to preconceived notions of what is and what is not reasonable. The theory must be put to the test by comparing its predictions with observed facts.

When this is done, the theory proves to be remarkably successful. For example, it suggests that the horizon problem has a simple solution. The regions on the opposite sides of the sky that emit identical amounts of microwave radiation *were* once in causal contact, according to the theory. If there was once a period of inflationary expansion, portions of the universe that were originally very close together would have become widely separated during the inflationary period. The smoothing-out process that made different parts of the universe alike could have taken place before the universe began to grow rapidly in size and to fling these regions apart so rapidly that they could never again come into contact.

The inflationary theory also suggests a solution to the flatness problem. Calculations indicate that at the end of the period of inflation, the universe was expanding at a rate very close to the critical velocity. The dynamics of the inflation process bring this about. The fact that the average curvature of space in the universe is close to zero is an automatic consequence of this. A rate of expansion near the critical value gives a mass density that is close to the dividing line too.

The theory is somewhat less successful in dealing with the galaxy formation problem. It does imply that there would have been fluctuations during the inflationary period that would have created regions of higher-than-average density. But the theory predicts conglomerations of matter that are much larger than galaxies. Although the calculated results do not quite correspond to the observed facts, cosmologists are nevertheless encouraged. A prediction of galaxies of the wrong size is better than a prediction of no galaxies at all.

The inflationary universe theory is a speculative theory that is built on other theories—the GUTs—that are themselves speculative. As a result, one cannot have complete confidence in the picture that it draws of events which took place in the early universe. However, it is hard to believe that a theory which explains so much does not contain a significant element of truth. The fact that the theory explains the appearance of the present-day universe so well suggests that some kind of inflationary expansion probably did take place.

There were some problems with the original version of the inflationary universe theory. In particular, it predicted that as the universe made the transition from the false vacuum state, it did not do so in a uniform manner. If this happened, the result would have been that bubbles of space-time in the true vacuum state would form at random, grow, and finally coalesce as their expansions caused them to collide with one another.

If the transition from the false vacuum had taken place in this manner, the present-day universe would contain irregularities that we could detect. It would also contain large numbers of monopoles. Magnetic monopoles are concentrations of mass-energy that are created when the universe makes the transition from one energy state to another. An irregular transition would have created quite a lot of them.

The "freezing" predicted by the original inflationary theory can be likened to the freezing of a lake in chunks. If a lake froze in this manner, an irregular surface would result. An isolated block of ice would be formed here, another there. One would be able to see where the frozen blocks had grown together, and the frozen lake would have a surface that contained a number of bumps and valleys.

A reasonable inflationary theory, on the other hand, would predict that the transition from the false vacuum state would resemble the freezing of real lakes. If detectable irregularities are not to be produced, the transition should be slow and gradual. It should resemble the kind of transition we see every winter, where a thin sheet of ice will form on a lake, and then gradually thicken as the temperature drops.

A modification of the inflationary universe theory was proposed by Guth in 1982. Called the *new inflationary scenario*, it avoids the difficulties encountered by its predecessor by making the assumption that the quantum fields in the early universe had such a character that the transition was a gradual one. In the new inflationary scenario, *domains*, or bubbles of space-time, are still created. However, they turn out to be much larger than the domains predicted by the older version of the theory. In particular, they turn out to be much larger than the observable universe. Thus if we really do live inside such a domain, its boundaries would be so far away that we could not tell they were there.

The new inflationary scenario (finally) solves the monopole problem. If the transition is relatively smooth, the concentrations of energy that give rise to monopoles are fewer. According to the new version of the theory, the number of monopoles in the universe would be so small that they would be difficult, or impossible, to detect.

If the reader gets the impression that a certain amount of "fiddling around" is necessary if one is to make the inflationary universe theory work, he is probably right. But there is nothing illegitimate about such a procedure. When physicists seek to penetrate the unknown, they can often do little at first except blunder around until they happen to hit upon the right ideas. The first theories to be formulated about a previously unexplained phenomenon often contain elements of an ad hoc character. Then, as time passes and scientists gain a better understanding of the phenomena that they are trying to describe, these first theories are replaced by others of a more coherent character.

For example, when physicists first tried to explain why they were unable to detect the presence of an ether, numerous hypotheses were invented. Some of these contained an element of truth. Lorentz, for example, correctly surmised that objects would appear to contract when they traveled at high velocities. Only after Einstein had propounded his special theory of relativity was the nature of this contraction really understood. However, this does not detract from Lorentz' achievement. He did manage to catch a glimpse of the truth that Einstein was later to see in its full clarity.

There is reason to believe that much of what the new inflationary theory tells us may be reasonably accurate. If it is, we may expect to gain an even better understanding of the events that took place in the early universe as theoretical work progresses. In particular, a theory which successfully unified all four forces might explain events in the early universe even more accurately, and possibly even provide us with some clues as to what might have been happening before the Planck time.

Of course, it is also possible that the inflationary theory will eventually be proved wrong. It may be that none of the GUTs upon which it is based is correct, that the goal of the unification of the forces will prove to be a chimera. It is possible that physicists have taken the wrong kind of approach, and that we will begin to understand the real nature of particles and fields only when some new Einstein hits upon a novel and unexpected idea that clears up a number of outstanding problems in a single stroke.

Although it is conceivable that such a thing may happen, most contemporary physicists do not consider it very likely. That the current approach is probably the right one is confirmed by the successes that have been obtained so far. Unification of the weak and electromagnetic forces has led to the prediction of the existence of new particles, which have been observed. Although the GUTs have not been experimentally verified, they can be used to clear up problems that have puzzled physicists for years, such as that of the apparent nonexistence of antimatter. Thus there is reason to suspect that physicists are on the right track and that, though a great deal of theoretical work remains to be done, the new inflationary scenario is at least approximately correct.

If it is, then there is nothing to stop us from attempting to take another step forward, carrying our speculation even further. If we provisionally assume that the inflationary theory gives a correct description of the universe, we can attempt to see what this might imply about the nature of matter and energy, of space-time, and of the universe itself.

As we have seen, the inflationary theory implies that our cosmos is made up of numerous "miniuniverses," bubbles of space-time that were filled with matter and energy during the period of inflationary expansion. Since the inflationary period ended, these bubbles have been growing into domains much larger than the observable universe. If the theory is correct, there could be an infinite number of such miniuniverses separated by domain walls of pure energy.

For that matter, there could be bubbles of space-time that will never come into physical contact with our universe. In other words, there could be other universes that we will never be able to observe or reach. If they exist, it might not even be possible to say "where" these universes are. After all, the word "where" refers to a location in space, and such universes would exist outside of our region of space-time.

If the universe grew from a tiny bubble of expanding space-time that was created by an unknown kind of quantum fluctuation of space-time, there is no reason why universes could not be created in this manner over and over again. After all, if something can happen once, it is reasonable to suspect that it can happen many times. There could be an infinite number of other universes.

Since we do not know how to speak of events which took place before the Planck time, we cannot say whether the universe was created in time, or whether time itself came into being when our bubble-universe popped into existence. In other words, there may have been no such thing as "before" the creation. On the other hand, time may extend into the infinite past.

Incidentally, these two possibilities—infinite time and time that had a beginning—were considered by philosophers long before modern physics began. The ancient Greek philosophers tended to believe that the world had always existed, and that it went through cycles of creation and destruction that repeated themselves endlessly. St. Augustine, on the other hand, taught that time had come into existence when God created the world. According to Augustine, there was no such thing as "before" the creation. It appears that some of the ideas of contemporary physics are not so new after all.

Yet another question that the inflationary universe theory raises is the following: How do we know that our universe is really in the true vacuum state? Just as an electron in an atom can exist in many different energy states, could there not be a number of different energy states of the vacuum? Presumably there could. And, if there were, we could not be certain that our universe went into the lowest one, the true vacuum state, when the inflationary expansion ended.

If this were the case, our vacuum could be unstable too. The universe could suddenly undergo another transition from one state to another. But if physicists ever found theoretical reasons for believing that this was true, they would not be able to confirm their hypothesis. For if the universe did undergo another energy transition, the change would propagate at the speed of light, and all the world's physicists would be vaporized before they would be able to observe that a transition had taken place.

In my view, the most interesting implication of the inflationary universe theory is not the possibility that we may all suddenly cease to exist. It is the suggestion that all of the matter and energy that we observe and, indeed, the universe itself may be nothing more than the result of a random fluctuation in quantum fields. The reason that I find the answer suggested by the inflationary theory so appealing is that it seems to answer the old philosophical question, Why is there something rather than nothing?

It may be that "something" and "nothing" are really not all that dissimilar. In fact, when we look at the present-day universe, it is difficult to find quantities that are significantly different from zero. Matter and energy seem to balance one another out, at least in a rough way. The average curvature of space is near zero. For all we know, it may be equal to zero exactly. The universe, as far as we can tell, has no net electrical charge. Matter, after all, is electrically neutral. Nor does there seem to be any net color charge. Free quarks are not observed; they seem to be present only as constituents of such colorless particles as baryons and mesons.

According to the general theory of relativity, the universe could conceivably be in a state of rotation. We could not say that the universe was rotating with respect to anything external to it, as the earth rotates with respect to the stars. However, such a rotation could be detected because it would alter the character of the microwave background radiation. But measurements indicate that if any such rotation exists, it is not significantly different from zero.

One is almost tempted to proclaim, Everything and nothing are really the same thing. However, it would probably be better not to become too caught up in such profound-sounding but meaningless slogans. It is possible that the universe came from nothing. Someday it may return to nothing. Many of its properties have a surprisingly nothinglike character. Nevertheless, it very obviously exists. It just so happens that the universe bears little resemblance to what we thought it was a century, or a decade, or even a few years ago.

Part Three

The Nature of Reality

7

The Knowable

The nineteenth-century positivist philosopher Auguste Comte once predicted that scientists would never know anything about the interiors of stars. He thought that since these regions could not be directly observed, they would be unyielding to scientific investigation.

Of course Comte was wrong. Today, the processes that take place in stellar interiors are understood very well. It is true that scientists can see into stars no more than they could in Comte's day. However, there are numerous ways to make inferences about stellar evolution and structure.

We know that nuclear reactions provide the energy that allows stars to radiate light and heat. Our sun has been shining for about five billion years. Only nuclear energy could allow it to remain bright for so long a time. Since the sun's diameter and mass are known, it is possible to calculate the pressures and temperatures that exist in its interior. It is possible to calculate the rates at which nuclear reactions would proceed under such conditions. Measurements made in nuclear physics laboratories on the earth provide information that can be applied to processes which take place in the center of the sun.

The quantities of energy radiated by the sun have been measured. Consequently, we know exactly how much energy these reactions produce. This provides a check on theory. If assumptions about conditions in the sun's interior were wrong, then there would be discrepancies between the calculated rates of energy production and the energy production that is measured.

Although the chemical composition of the sun's interior cannot be determined directly, it is possible to measure the amounts of the various chemical elements that are present in the sun's atmosphere. As the light from the interior radiates outward through the hot gases that make up the solar atmosphere, certain wavelengths of light are absorbed. Each chemical element has its own characteristic "signature." Thus it is necessary only to observe the missing wavelengths to determine what elements are present. In fact, one element, helium, was observed on the sun, in 1868, long before it was found on the earth. The name "helium," by the way, comes from *Helios*, the sun god of Greek mythology.

The other stars are much farther away than the sun; the nearest lies at a distance of 25 trillion miles. Nevertheless, astronomical observations have provided a tremendous amount of valuable data about them also. Although it is not possible to study them in as much detail as the sun, numerous different kinds of stars can be observed. Astronomers have studied stars that are much larger than the sun, and stars that are much smaller. They have observed stars in all the different stages of stellar evolution. They have seen stars which are only now being created, and stars that exhausted their nuclear fuel billions of years ago. Astronomers can make out not only middle-aged stars like the sun, but also red giant stars that are going through their death throes, and white dwarfs that burned out long ago but which continue to glow dimly only because they have not yet radiated away all their residual heat.

One can assert that astronomers and astrophysicists not only have a good understanding of stellar structure, but can also describe the stages of stellar evolution and predict what a given star will look like billions of years from now. In fact, stellar interiors are understood somewhat better than the interior of the earth. It is hard to imagine a scientific prediction that has turned out to be more inaccurate than Comte's.

However, his prediction does raise an interesting question, or rather a series of related questions: Is there anything which is unknowable in principle? Or should we expect that physicists will eventually discover everything there is to know about the universe? Are there limits to human knowledge? Are there limits to human comprehension? Does science ever discover that certain things cannot be known?

It is not easy to answer such questions, for ideas about what is, or is not, knowable have changed as dramatically as conceptions of physical reality. Nowadays scientists commonly speculate about questions that were once considered to be "beyond physics." At the same time, they have encountered limitations unlike those experienced by previous generations of scientists. An example of this was pointed out in a previous chapter: If physicists are to have any hope of experimentally confirming some of the unified theories that are being proposed, they must find ways to understand processes which take place at energies far higher than those which can conceivably be produced in particle accelerators.

A good example of the way in which ideas about what is knowable have changed is given by Steven Weinberg in his book *The First Three Minutes*. As Weinberg points out, the cosmic microwave radiation background was discovered only in 1965, even though the technology needed to observe it existed in the 1940s and 50s. Furthermore, although the existence of the microwave background had been predicted in 1948, most of the astronomers and astrophysicists of the 1960s were unaware of this fact. Most of them did not know that the radiation should exist, or that an important confirmation of the big bang theory could be obtained if it was observed.

As a result, when Penzias and Wilson did discover this background, they made their discovery by accident. When they first heard this cosmic "static" on their radio telescope, they thought that the source of it must be something in their antenna or in the electric circuits connected to it. It was only after they had eliminated all other possibilities, and had consulted other scientists about their "problem," that they realized that the radio noise they were hearing was coming from the universe.

The fact that significant scientific discoveries are often made by accident is well known. Scientific knowledge has been advanced on numerous occasions when scientists have stumbled upon new and unexpected phenomena. However, the discovery that Penzias and Wilson made does not fit the usual pattern. The phenomenon that they discovered was one that had been predicted seventeen years before, and then forgotten.

Weinberg suggests that there may have been several reasons for the existence of such a surprising state of affairs. He cites certain theoretical problems that afflicted the big bang theory when it was first proposed. He points out that there may have been a breakdown in communication between theorists and experimental scientists. The experimentalists apparently didn't know the background radiation should be there, while the few theorists who did know that it might exist didn't realize that it could be detected.

But this was not all there was to it. According to Weinberg, there was a third reason why no one had set out to look for the background radiation. He expresses the opinion that the physicists of the day generally did not take any theory of the early universe very seriously. Although the big bang theory could be used to make reasonably precise predictions about phenomena that could be observed in the universe, few scientists followed where the theory led. As Weinberg points out, "It is always hard to realize that these numbers and equations we play with at our desks have something to do with the real world."

After the discovery of the cosmic microwave background became known, however, the big bang theory quickly came to be considered something more than a theoretical fantasy. Scientists realized that they could make observations of the present-day universe and use them to make deductions about events which took place when the universe was young. They found that the microwave background was a kind of afterglow of the big bang. Measurements of the helium and deuterium concentrations in the universe allowed them to describe events that had taken place when the universe was only a few minutes old.

The deductions that are made about conditions in the big bang fireball are analogous to those which are made about the interiors of stars. One can no more travel 15 billion years back in time in order to observe the universe than one can travel into the interior of a star in order to measure quantities like temperature and pressure directly. However, in either case, one can make inferences about the events that take place.

As physicists discovered that it was possible to put together a reasonable theoretical picture of the universe during the first few minutes of its evolution, they began to speculate about events which happened even earlier. They began to speak of the conditions that existed during the first few seconds, then the first few hundredths of a second; finally, they began to attempt to look back all the way to the Planck time. The change in outlook is striking. In the 1960s, many physicists doubted that it was even very meaningful to speak of the big bang as an event that really took place. Today, scientists speak of events which took place during the first 10^{-32} seconds.

As scientific knowledge has increased, ideas about what is knowable have changed dramatically. At one time, such questions as "What happened before the big bang?" and "What caused the big bang?" were considered to be almost metaphysical in nature. It was said that such things were "beyond physics." Today, some scientists are beginning to speculate about such questions, to attempt to go back even further than the events described by the inflationary universe theory.

There is not yet any theory that will allow scientists to describe processes that took place before the Planck time. In order to do so, it would be necessary to develop a theory of quantum gravity. At present, no one even knows what such a theory would have to be like. However, the lack of such a theory has not proved to be an impenetrable barrier to speculation.

Since it is known that space and time would be subject to quantum fluctuations of some kind during the first 10^{-43} seconds, some physicists have attempted to see if anything can be said about the character that such fluctuations might have. Others have begun to ask if it might not be possible to ask what happened before the big bang after all. Though they don't yet have any way of knowing what the correct answer might be, they have found that it is at least possible to enumerate the possibilities.

As I have mentioned previously, there may have been no "before." Time may have been created in the quantum fluctuation that brought the universe into existence. Alternatively, it is possible that time existed before the big bang while space did not. There are yet other possibilities. The big bang may have been a "bounce," not a bang. The universe may have previously been in a state of contraction, and quantum effects of an unknown nature might have caused this contraction to reverse itself. Alternatively, our universe may have been created when a small region of space-time split off from a "mother universe." It has been suggested that it might be possible for universes to reproduce themselves in such a way. Initially, mother and daughter universes would presumably be connected by a kind of space-time bridge, which would soon dissolve, leaving the two spatially disconnected.

No one knows exactly how the universe was created. However, it has been discovered that it is at least possible to ask. The discovery of answers to old problems leads inevitably to new questions. The more we know about the universe, and the early universe in particular, the greater the number of new queries that can be posed. Everyone is aware that the making of discoveries is an important part of scientific progress. Relatively few realize that progress can also be made by inventing new questions to ask. The more we know, the more we realize how much there is that might some day be known.

Similar changes in ideas about what is knowable can be seen in the field of quantum theory. During the early 1930s, physicists knew of only three particles, the electron, the neutron, and the proton. They did not ask why the charges of the proton and the electron should be equal and opposite, or why the particles should possess the particular masses that they were observed to have. It was generally believed that it was the task of physics to describe observed phenomena only. Few scientists asked why the laws of nature should have the particular character that they did, or why physical constants had certain values and not others.

For example, if the proton was observed to be 1836 times as heavy as the electron, there was nothing more to say. If one established that the electromagnetic force was so many times stronger than the gravitational force, there was nothing to add. If it was found that electrons had spins of a certain magnitude, one did not ask why the spin was not larger or smaller. The question, "Why do we observe these magnitudes, and not others?" would have been thought to be meaningless.

Nowadays, physicists realize that these questions may not be meaningless after all. One of the reasons for seeking a unified theory of the forces is that such a theory might specify exactly what particles should exist and what properties they should have, including mass, electric charge, and spin. Such a theory might give reasons why a proton should be 1836 times as heavy as an electron, and tell us why charged particles always have a particular amount of charge.

In particular, supergravity theories predict the existence of certain particles with certain properties. For example, the supergravity theory that has received the most theoretical attention predicts that there should be exactly 163 different subatomic particles, and divides them into subgroups according to the amount of spin that they possess. It is by no means certain that the particles predicted by the theory correspond to those that are actually observed. In fact, there are particles whose existence the theory does not seem to be able to account for. However, a better theory might be able to, and to explain all their properties as well.

At the moment, there are numerous unanswered questions about the nature of fields and of particles, and no one is really sure what kind of theory is most likely to be correct. However, if some supergravity theory, or superstring theory, or some theory that is based on an approach that no one has yet thought of, proves to be successful, there is a chance that, for all practical purposes, physics will come to an end. Such a theory might give us full knowledge about the quantum fields that exist in our universe. If this knowledge is attained, we will—at least in principle—know everything there is to know about the nature of physical reality. All that remains will be to work out some of the details, such as the applications of the theory to processes that took place in the early universe.

Or, at least, this is the hope that was expressed by the British physicist Stephen Hawking in his inaugural lecture as Lucasian professor of mathematics at Cambridge University in 1980. Entitling his lecture "Is the End in Sight for Theoretical Physics?" Hawking suggested that supergravity theory might fully explain the nature of the physical world.

If physicists found such an ultimate theory, they might discover that all their questions had finally been answered. They might discover that there is only one kind of universe that is logically possible, that this ultimate theory would demonstrate that physical reality is the way it is because anything else would imply contradictions.

Einstein seems to have thought along similar lines at certain times in his life. On one occasion he said that one of his reasons for working on physics was to find out whether or not God had any choice in creating the universe the way He did. Einstein, who did not believe in the personal God of Judaism and Christianity, often spoke of "God" when he wanted to make statements about the natural order. He once explained this by saying that his God was one who revealed Himself "in the harmony of all that exists." Thus it is possible to paraphrase Einstein's statement by saying that he wanted to find out if the universe was the way it was because nothing else would be conceivable.

The idea that only certain kinds of physical laws are logically possible and the idea that we will eventually discover a theory that explains everything are clearly not the same. However, they are related. An all-embracing theory, after all, might be recognized for what it was because it had become apparent that it was the only possible theory.

Some physicists take issue with Hawking's idea that a complete theory will eventually be found. For example, John Archibald Wheeler says, "I admire Hawking, but I can't agree with the idea that there's any magic *equation!*" (his italics). The physicist Freeman Dyson of Princeton's Institute for Advanced Study also disagrees. Dyson has commented, "If it should turn out that the whole of physical reality can be described by a finite set of equations, I would be very disappointed. I would feel that the Creator had been uncharacteristically lacking in imagination."

Dyson says that he suspects that the laws of physics are inexhaustible. To support this conjecture, he has given an argument that is based on a famous theorem in mathematics, known as *Gödel's proof.* This theorem, which was stated and proved in 1931 by the Czech mathematician Kurt Gödel, states that it is not possible to prove that any mathematical system at least as complex as ordinary arithmetic is consistent. There is no way to demonstrate, in other words, that such a system will not eventually become bogged down in contradictions. Furthermore, if such a system is consistent, then it must be *incomplete*; there must exist true statements which cannot be proved within the system. Such statements are called *undecidable*.

Examples of undecidable statements are known. However, in order to give examples of such statements, it would be necessary to go on a long digression and to explain some rather technical ideas in modern mathematics. In order to avoid doing this, I propose to engage in a bit of cheating, and to use as an example a statement that mathematicians only *suspect* is undecidable.

There are many conjectures in mathematics which mathematicians think are true, but which have never been proved as theorems. One of the most famous, known as Goldbach's conjecture, was made in 1742 by an amateur German mathematician named Christian Goldbach. Goldbach happened to notice that every even number could apparently be expressed as the sum of two primes.* For example,

$$4 = 3 + 1$$
$$6 = 5 + 1$$
$$28 = 23 + 5$$
$$100 = 47 + 53$$

Goldbach could find no exceptions to this rule. To be sure, there were some even numbers that could be expressed as the sum of primes in more than one way (for example, 28 is also 11 + 17). However, as far as he could tell, there did not exist any even numbers which could not be broken down in this manner.

One might expect that such a simple idea would be easy to prove as a mathematical theorem. However, Goldbach was unable to find a proof. So he wrote a letter about the problem to the Swiss mathematician Leonard Euler. Although Euler was probably the

---

* A prime is a number that is divisible only by itself and by 1. For example, 1, 2, 3, 5, 7, and 11 are primes.

best mathematician in the world at the time, he was unable to prove the conjecture either. The problem is still unsolved today. The best that modern mathematicians have been able to do is to show that every even number can be expressed as the sum of no more than 300,000 primes.

No counterexample to Goldbach's conjecture has ever been found. The conjecture has proved to be valid for every even number that anyone has ever tested. But showing that a statement is true in particular cases is not a mathematical proof. In order to prove the conjecture as a theorem, it would be necessary to show that it is true for every even number, including all those that are much too large for anyone to write down. After all, a single exception would be sufficient to prove it wrong.

Nowadays most mathematicians believe that the conjecture is true. But they do not expend much effort trying to prove it. Many of them suspect that proof is impossible, that Goldbach's conjecture is an undecidable statement.

It is often said that Gödel's theorem implies that ultimate mathematical truth can never be attained. This is true to a certain extent. Unless one deals with mathematical systems that are so simple that they are not very interesting, one has to contend with the fact that inconsistencies may eventually be discovered, or that undecidable statements will crop up. However, there is another way of looking at things. As Dyson points out, one could equally well say that Gödel proved that mathematics is inexhaustible.

Whenever an undecidable statement is found, one is free either to assume it as an axiom, or to assume that it is false. If the statement can be neither proved nor disproved, either assumption would be consistent with the rest of the mathematical system. Since the two assumptions will lead to different kinds of mathematical results, one has two slightly different mathematical systems where only one existed before. It appears that we have not just one possible kind of mathematics, but many.

For example, there is a statement, part of the mathematical theory of infinite numbers, that is known as the *continuum hypothesis* and has been shown to be undecidable. Consequently the continuum hypothesis may be either used as an axiom, or denied in any of a number of different ways. As a result, one mathematical theory has branched off into several. Naturally, all of the resulting theories are equally "true," even though they lead to different conclusions.

In Dyson's opinion, Gödel's theorem implies that mathematics will always have "fresh questions to ask and fresh ideas to discover." He hopes that it will be possible to prove that the world of physics is just as inexhaustible.

There is nothing in physics that is the analogue of Gödel's theorem. However, the idea that physics has its undecidable statements too sounds quite reasonable. After all, mathematics and theoretical physics are similar in many respects. Mathematicians begin by assuming certain axioms and postulates, which they use to prove theorems. Physicists construct their theories by making what seem to be reasonable assumptions. They put these assumptions in mathematical form, and try to see what consequences can be deduced.

Of course, there is a respect in which mathematics and physics are very different. Mathematics is not an empirical science. Mathematicians do not perform experiments to test the correctness of their conclusions. Such an idea would be absurd. Mathematics, after all, is the science of finding logical consequences. If one knows that certain axioms imply certain theorems, that is all that there is to know.

The British philosopher Bertrand Russell once defined mathematics as "the subject in which we never know what we are talking about, nor whether what we are saying is true." Russell's purpose in making this somewhat paradoxical-sounding statement was to emphasize the nonempirical nature of mathematics. Even when neither axioms nor theorems seem to have anything to do with the real, observable world, one can still use mathematical techniques to show that if certain axioms are assumed, certain consequences logically follow.

Even though theoretical physics has a highly mathematical character, physicists proceed in a very different way. They are not free to assume anything they like. They must use assumptions which are physically reasonable. If these assumptions lead to conclusions that are contradicted by experiment, they must be discarded, no matter how appealing they might have seemed. It is possible that the interaction between theory and experiment has the effect of eliminating undecidable statements from physics, or at least some of them.

It should be emphasized that no one really has any idea whether physics contains undecidable statements or not. However, it is really not very difficult to think of theoretical hypotheses that might have that character. For example, there is the idea that there may exist other universes that have no physical connection to our own. If such universes exist, they cannot be observed. Thus their existence could be neither proved nor disproved.

Nowadays, many different variations of the "other universes" idea can be found in physics. Most of these are offhand, speculative suggestions. However, there is one particular "other universe" theory which has been developed in great detail. This is the *many-worlds interpretation* of quantum mechanics.

Quantum mechanics is one of the most successful theories that has ever been discovered. It is the foundation of virtually all of modern physics, and the predictions that it makes have been confirmed to an unprecedented degree of accuracy. However, even though quantum mechanics can be used to make precise numerical predictions, and even though other successful theories can be built on it, it is not an easy theory to interpret. It turns out to be very difficult to say exactly what the theory means. The picture that it gives us of the physical world is sometimes rather ambiguous.

For example, the uncertainty principle tells us that the position and momentum of a particle cannot be measured simultaneously. The more accurately the position is determined, the less one can say about a particle's momentum. Or, if the momentum is found with great precision, then the particle's position is unknown.

The puzzling thing about all this is that one can choose to measure either position or momentum. In principle, either could be determined to any required degree of accuracy. Thus it is not easy to understand the implications of the fact that we cannot simultaneously know both. It appears as though the experimenter's choice determines what properties a particle can have.*

---

* The implications of this will be fully discussed in Chapter 9.

Naturally, an experimenter is not required to measure either the position or the momentum exactly. He can perform a measurement that gives him partial information about both quantities. The uncertainty principle does not say that either one or the other must be determined; it says that if we multiply the uncertainty in the momentum by the uncertainty in the position, the product of the two numbers will always be about the same. In other words, one can perform an experiment which provides information about both position and momentum if one is willing to accept a result that leaves both quantities a bit "fuzzy."

Under such circumstances, the best that one can do is to represent both momentum and position by probabilities. One cannot say, for example, that the particle was seen in a definite place, or that it will occupy a definite position at some future time. The partial information about its position only gives the probability that it is within a certain distance of a particular point.

The probabilistic character of quantum mechanics exhibits itself in many other ways. For example, if a photon of light is made to collide with an atom, one cannot predict what will happen. It is only possible to speak of the probability that the energy of the photon will or will not be absorbed by the atom. Nor can one say when the atom will release this energy, reemitting a photon of light. It is possible only to speak of the probability that this will happen in a given period of time. In some cases, the atom may not reemit light of the same wavelength; it may give back the energy in steps. This is a matter of probability too.

Similarly, it is not possible to predict when a radioactive atom will decay (by emitting an alpha particle, for example). It is possible only to speak of the probability that this will happen within a certain time. There may be a certain chance that the decay will take place within the next ten seconds. There will be a much larger probability that it will take place during the next 10,000 years. But if one wants to say exactly when the decay will happen, the best that one can do is to make a guess.

If quantum mechanics cannot tell us when an event will take place, and if it sometimes cannot even tell us what will happen, one might think that it would not be a very useful theory. But this is not the case. Quantum mechanics can be used to make very accurate predictions when large numbers of particles, or of atoms, are involved. In any macroscopic quantity of matter there are many billions of billions of atoms. In such a case, the uncertainties will average out. For example, an individual atom in a fluorescent light tube may emit a photon at any time. But if there are countless billions of atoms in the tube, photons will be emitted at a steady rate. Similarly, even though one cannot predict when a given radioactive atom will decay, it is possible to define a *half-life* if the number of atoms is large. The half-life is just the time that will elapse before exactly half of the atoms decay. It is also the time for which the probability of decay of any atom is exactly 50 percent.

Perhaps it is fortunate that quantum probabilities average out in the behavior of everyday objects. Life would be even more unpredictable than it is if they didn't. In such a case, we would have to worry about whether or not light bulbs would steadily give off light. Television pictures would be subject to random disruptions. If the quantum uncertainties were large enough, we might even find that ice cubes sometimes melted when we put them in the freezer compartment of the refrigerator.

When the existence of quantum probabilities was discovered in the mid-1920s, many physicists found the idea disturbing. Einstein, in particular, would never accept the idea. Arguing that "God does not play dice," he made one objection after another to the concept. In Einstein's view, the fact that probabilities had to be used to describe the behavior of atoms and subatomic objects only reflected our lack of knowledge about their behavior. In his opinion, quantum mechanics was not a complete theory. It would eventually be replaced by a subquantum theory that could be used to predict the behavior of subatomic objects exactly. When this happened, probabilistic descriptions could be discarded.

Such a deterministic version of quantum mechanics would be called a *hidden-variable* theory. Hidden variables are quantities that could be used to make exact predictions. For example, if position and momentum could be known exactly, they would be hidden variables. It is always possible to plot the future trajectory of any object if one knows its present position and momentum and the forces that are acting upon it. Alternative, hidden variables could be quantities that could be used to determine position and momentum, or used to determine such quantities as the exact time at which an atom will emit an alpha particle. In a hidden variable theory, everything is predictable if we have enough information about an object. Probabilities need only be used when our knowledge is incomplete.

Over the years, a great deal of theoretical work has been done on the question of whether such a theory would be possible. This work culminated in a theorem proved in 1965 by John Bell, a theoretical physicist at the Center for European Nuclear Research (CERN), near Geneva. Bell showed that if one assumed that a local hidden-variable theory was possible, a certain mathematical formula—called *Bell's inequality*—could be derived.

The term "local" is an important one here. The idea that there can be no such thing as a causal influence or signal that travels faster than light is called the assumption of *locality*. If the assumption were false, then it would be possible for events to be instantaneously influenced by other events that took place great distances away. For that matter, it would be possible to send signals backwards in time. According to special relativity, a faster-than-light signal will seem to some observers to arrive at its destination before it is sent.

The assumption of locality is one that few physicists would care to give up. Special relativity is one of the foundations upon which modern physics is built. Its predictions have been tested over and over again. If it implies that the assumption of locality is true, that is a result that one should accept unless there is overwhelming evidence to the contrary.

The mathematical results that Bell obtained can be tested experimentally. The relevant experiments have been performed on a number of occasions, and it has been shown that Bell's inequality is violated in certain kinds of experiments with quantum particles. Since the standard probabilistic version of quantum mechanics is consistent with this result, while the assumption of local hidden variables is not, it appears that Einstein's idea that a deterministic subquantum theory might be found is untenable.

Unless one wants to accept the idea that nonlocal hidden variables might exist, it is necessary to conclude that the behavior of quantum particles is inherently indeterministic. The reason that we cannot predict when a radioactive atom will decay is not that we lack a complete theory. Quantum mechanics is complete; it tells us everything that it is possible to know about the atom. Decay cannot be predicted. Events in the quantum world are governed by chance.

It should be noted that quantum "probability" is not the same as the "probability" that is exhibited by the behavior of macroscopic objects. For example, we think of roulette as a game of chance. However, there is no inherent indeterminism in the game. There is no reason why it should be impossible to predict what number will come up on every spin of the wheel. If one could obtain information on the velocity at which the wheel was spun, the force with which the ball was propelled around the rim, and so on, the element of chance in roulette would disappear. It would only be necessary to feed the data into a suitably programmed computer, and one could win every time.

Quantum events have a different character. If one collected all the relevant data about a radioactive nucleus, it would not be possible to predict when the decay would take place, or in what direction the emitted particle would emerge. When we say that quantum events are governed by chance, we mean that a particle will emerge from an atom at a particular time and in a particular direction for no reason at all.*

Or at least this is the most common interpretation of quantum indeterminism. However, there is an alternative: the many-worlds interpretation of quantum mechanics, which was proposed by Hugh Everett III in his Princeton doctoral dissertation in 1957. According to this interpretation, there is no element of chance in quantum mechanics at all. A radioactive atom does not decay at some particular time by chance. On the contrary, it decays at *every* instant of time. And every time that it does, the universe splits in two.

This bifurcation presumably takes place every time that a quantum "choice" is made. This process creates a universe in which the atom decays after ten seconds, a universe in which it

---

* This does not imply that radioactive decay is "uncaused." Radioactive atoms decay because the particles within the nucleus have so much energy that, sooner or later, one of them will escape. Quantum mechanics tells us how such a decay takes place, though it cannot say when.

decays after three days have passed, a universe in which it decays after ten thousand years, and so on. At every moment, a new universe is created. This universe is identical to our own in every respect except that the atom decays at a different moment of time.

Naturally, every other quantum process that takes place causes additional splittings. Bifurcations of the universe are created every time an atom in the Andromeda galaxy emits, or does not emit, a photon of light, every time that an electron in intergalactic space encounters another particle, every time that a meson decays in a galaxy ten billion light-years away.

Since our bodies are part of the universes thus created, we split too. At every instant, countless, possibly infinite, numbers of duplicates of us are created. Yet this splitting process, if it occurs, would be inobservable; there is no way that we could communicate with our counterparts in other universes. In fact, since the universes created by quantum bifurcations would not be connected with one another in any way, there would be no way to verify that they really existed.

It all sounds very fantastic and quite unbelievable. And yet, there is no way that the many-worlds interpretation could be disproved. Since it is only an interpretation of the theory, accepting such ideas does not alter the numerical predictions that are made by the theory. There is no experiment that could distinguish between the many-worlds interpretation and the standard one. No observation is possible which would enable us to decide whether quantum events are truly random, or whether *all* quantum probabilities become reality in one universe or another.

If the theory is correct, then all possibilities are realized somewhere. There are universes in which the earth does not exist, and universes which do not contain intelligent life. There are universes in which the South won the Civil War, and universes in which the Roman Empire still exists. After all, a single quantum event could have a significant impact on history. For example, if a cosmic ray had altered the DNA of a single sperm cell of Philip II, king of Macedon, then Philip's son Alexander could have inherited a birth defect that would have caused him to die in infancy. In that case, Alexander would not have become Alexander the Great, and the world might be very different today.

Perhaps this example seems farfetched. However, it really isn't. Most quantum events have no observable effect on events in the everyday world. The ones that could conceivably affect the course of history are few indeed. But such quantum events can take place. And if all possibilities are realized, they will take place in one universe or another. Thus, if the many-worlds interpretation is correct, there are countless universes in which Alexander died in infancy, and countless others in which he was not even born.

Some critics of the many-worlds interpretation have objected that the idea of "alternate" universes that can never be observed is meaningless. In their view, the only kind of reality of which we can meaningfully speak is that which can actually be observed.

This is a perfectly legitimate objection. However, it is a philosophical criticism of the many-worlds interpretation, not a scientific one. The argument is forceful, but not absolutely convincing. Many philosophers and scientists find the alternative—the idea of quantum randomness—to be philosophically unpalatable also. They are unwilling to accept the idea that the universe is governed by chance.

Many of the objections to the many-worlds theory have been of a philosophical nature. Sometimes the objections are little more than expressions of philosophical discomfort with the ideas involved. For example, John Archibald Wheeler, who was one of the early supporters of the many-worlds interpretation, has given up on it because he has concluded that it "creates too great a load of metaphysical baggage." Although some critics have attempted to frame scientific objections to the theory, their arguments have not been altogether convincing. At the moment, it appears that there is no valid scientific way to determine whether or not the interpretation is valid.

There is a remarkable similarity between the ideas on which the many-worlds interpretation is based and undecidable statements in mathematics. Existing scientific theory can be used neither to prove nor disprove the idea that these other universes may exist. The question of their existence may be one that we will never be able to answer.

Naturally, we cannot call the assertion "The many-worlds interpretation of quantum mechanics is correct" an undecidable statement. There may be ways in which the analogy between it and some of the unprovable statements that are encountered in mathematics breaks down. Nevertheless, the fact that such apparently unanswerable questions exist would tend to confirm Freeman Dyson's belief that physics, like mathematics, is inexhaustible.

The existence of such unanswerable questions does not necessarily disprove Hawking's suggestion that physics may soon be able to explain "everything." After all, there is no reason why we can't take "everything" to mean all the physical phenomena that we can actually observe. Universes that cannot be detected are obviously not in this category. Apparently, it is indeed possible that physics may come to an end in our lifetime.

It is possible, but I suspect that it will not happen. Perhaps it is a bit presumptuous of me to insert my own opinion into a discussion of those of physicists of the stature of Hawking and Wheeler. However, I can't refrain from observing that this is not the first time that the idea that physics was near an end has been put forward.

In 1875, when Max Planck was seventeen years old and preparing to enter the university, he approached the head of the physics department at the institution that he planned to attend. Planck wanted to discuss his future course of study. He did not find the department head to be very encouraging. The latter advised Planck to study some other subject instead, assuring the young man that all the important discoveries in physics had already been made. "Physics is a branch of knowledge that is just about complete," he is supposed to have said.

There may be some apocryphal elements in this tale. Similar stories have been told of other professors and of other young students who were later to become renowned physicists. However, it is possible that all of these stories contain elements of truth. During the latter part of the nineteenth century, it was commonly believed that the tasks of physics were nearly complete.

At the time, it seemed that there was little left to be discovered. Newton's laws of motion and his law of gravitation appeared to provide the last word on those subjects. Scientists understood such phenomena as heat, sound, electricity, and magnetism. Maxwell had successfully combined explanations of the latter two in a unified theory, which also explained the nature of radiation. It is not surprising that many scientists believed that it only remained to work out explanations for a few remaining minor puzzles.

In the early 1930s, similar views were expressed. By this time, quantum mechanics had proved to be a successful theory of sub-atomic phenomena. Dirac's relativistic theory had cleared up the mysteries about such phenomena as electron spin. The three basic constituents of matter—the electron, the proton, and the neutron—had been discovered, and were reasonably well understood. When the German physicist Max Born assured his colleagues that physics would be over in six months, not all of them agreed. However, it must have seemed that Born had a point.

Physics didn't come to an end in the years following 1875, nor did this happen in the 1930s. It didn't because some of the little unsolved scientific puzzles that were supposedly soon to be cleared up turned into big problems. And when they did, whole new fields of physics were opened up, and revolutionary new theories were developed. On both occasions, scientists discovered that they understood much less than they had thought, that their knowledge of the universe was nowhere near complete.

I suspect that something similar may happen in the years ahead. In the past, scientific discovery has always led to further discovery. When a question has finally been answered, new and more profound questions have generally arisen in its place. I admit that the patterns we discern in the past are not always a reliable guide to the future. So it is possible that Hawking may be right when he predicts the end of physics. However, the similarity between this prediction and those that have been made in the past should at least give one pause for thought.

It may be that the point of view that one adopts depends upon one's philosophical outlook. If one believes, as Hawking apparently does, that the universe is ultimately comprehensible, then it seems natural to believe that physics will one day be complete. If, on the other hand, one believes that there is a sense in which the universe will always be beyond complete human comprehension, that there will always remain a residue of unexplained—and possibly inexplicable—phenomena, then one is likely to conclude, like Dyson, that physics is inexhaustible or, like Wheeler, that there is "no magic *equation*." After all, the fundamental question is

not "Will physics ever be complete?" but rather, "Is an understanding of the ultimate nature of physical reality something that
is within the reach of the human mind?" In inclining to the latter,
it is not my purpose to denigrate human intelligence. I simply feel
that there is more to the universe than we are able to imagine.

8

Truth

Just how seriously should one take the theories of modern physics? Given the fact that theories are constantly being overthrown and replaced by new theories, to what extent can we say that the findings of modern physics really represent the "truth"? If we accept the picture of reality that is implied by a theory that is accepted today, is there not a chance that the picture will be proved wrong tomorrow?

Just as scientific conceptions of reality have changed, the answers to these questions have changed also. In Newton's day, scientists generally believed that they were discovering the laws of nature that had been used by God in the construction of the universe. In their view, the very existence of physical law implied that there must be a Lawgiver. According to this line of reasoning, it was possible to gain a glimpse into eternal truths by decoding natural phenomena.

It was believed, for example, that Newton's inverse-square law of gravitation was exact. Perhaps, as Newton believed, the law of gravitation was not sufficient to keep the solar system in perfect working order. God, Newton suggested, might occasionally have to intervene, in the manner that a clockmaker might adjust his clocks, to keep planets from drifting into inappropriate orbits and from colliding. However, the gravitational law itself remained unchangeable; it was one of the tools that God had used in constructing the universe. As such, it was perfect and exact.

As time went on, this outlook began to change. By the end of the nineteenth century, physicists were thinking of scientific theories as approximations that were only models of reality. Theories bore the same relation to the physical world that an architect's model did to a building, or that a detailed drawing did to a scene in nature. The correspondence between model and reality might be very close, but it could never be exact.

This also remained the prevailing outlook during the twentieth century. It was felt that it is the task of physics not to find exact laws, but to obtain better and better approximations. As these approximations have improved, the fit between theory and reality has naturally become better. However, according to this view, it can never be perfect.

When Einstein's general theory of relativity replaced Newton's inverse-square law of gravitation, the old theory was not really "disproved." All that happened was that a good approximation was replaced by a better one. Furthermore, there is reason to expect that even better approximations will be found in the future. If quantum mechanics and general relativity can somehow be combined, we would have an even more accurate theory, one that might describe the structure of space-time in the Planck region, where general relativity is expected to fail.

Until recently, virtually all modern physicists thought of theories as models. Indeed, this seems to be quite a reasonable point of view. After all, the entities with which modern physics deals often cannot be observed directly. As a result, they must be regarded as mathematical constructs.

For example, no one has ever seen an electron. To be sure, we can observe a spot of light when an electron strikes a fluorescent screen. But we cannot see the electron itself. We can only infer its existence from various kinds of observed phenomena. Similarly, one cannot see gravitational fields, or the curvature of space-time.

It is possible only to calculate the consequences that follow from the assumption that these things exist, and then to compare theoretical predictions with experiment.

This point can be illustrated by a look at the methods that physicists use to detect subatomic particles and to study their interactions. There are a number of methods currently in use. One of the most common is to allow beams of particles to pass through a *bubble chamber*, a large tank that has been filled with liquid hydrogen. As the particles pass through the tank, they create tiny bubbles of hydrogen gas. After these bubbles have been photographed, experimental physicists can study them at their leisure.

Physicists often speak of "seeing" the particles that produce tracks in bubble chambers. In reality, all that they ever see are strings of hydrogen bubbles. The idea that these strings of bubbles were produced by objects such as baryons, leptons, and mesons is an inference.

It is the task of theoretical physicists to make up theories about the behavior of these objects. But if theoretical predictions are confirmed by experiment, one still has not demonstrated that they really exist. All that one can say is that the model leads to results that correspond reasonably well to reality. Naturally this does not demonstrate that the assumptions that have been made are the only ones possible, or that better assumptions cannot be found.

After all, the fit between theory and experiment is never perfect. There are often unexplained discrepancies. Sometimes matters are even worse. In every field of physics, there has always been a body of experimental results that have failed to correspond to theoretical predictions at all. These anomalous results are often ignored, on the theory that there may have been something wrong with the experiments that produced them, or that unknown factors might be influencing the results. Indeed, this often turns out to be the case. On the other hand, it is not hard to cite "small anomalies" which were never eliminated, which were explained only when revolutionary new theories were finally developed. An example of this would be the anomaly of 43 seconds of arc per century in the orbit of Mercury that could not be explained until the advent of the general theory of relativity.

Perhaps it would be illuminating to go back to the analogy according to which theoretical physics is like an architect's model of a building. Naturally, if we want to pursue the analogy very far, we probably ought to imagine that the usual procedure has been reversed, that the building existed first and that the model was constructed afterwards. It is not necessary to say how the building came into existence. Some would say that it was created by God. Others would prefer to think it is the result of certain natural processes (such as the creation of the universe from a quantum fluctuation). But it really doesn't make any difference which view is correct. It is obvious, in any case, that the building is there.

Most of the time, physicists labor to improve their model of the building. They add a few touches here and a few there in order to make the correspondence between theory and reality more complete. But on some occasions they discover that something in their model looks a bit askew. At first they will try to ignore this. However, in many cases, the defect will grow more and more obvious and annoying. When this happens, they sometimes have to dismantle parts of the model and rebuild them anew. On rare occasions, they discover that there is a problem in the foundations. When this happens, they often have to take the entire model apart and start over again.

Whether they are only elaborating upon their model or rebuilding it completely, their goal is always the same: to make it resemble the building more and more exactly. Indeed, at every step, the resemblance does grow better and better. Eventually it is so good that they begin to ask if the correspondence between building and model will ever be perfect.

This is one of the questions that we find being asked today. So many improvements have been made in physicists' models of reality that some scientists have begun to ask whether it is possible that their models are approaching perfection. Some of them have drifted back toward the point of view that was characteristic of the age of Newton, that it is possible to discover exact laws of nature.

The search for a theory of supergravity, in particular, has evoked this point of view. A successful supergravity theory would explain so much, and answer so many questions, that some physicists have expressed the opinion that supergravity *is* reality.* As we have seen, this point of view has been championed by Hawking.

Other physicists argue that models can never be perfect, or even particularly accurate, representations of the physical world. Wheeler, for example, questions that we can ever know "reality." "What we call 'reality,'" he says, "consists of an elaborate papier-mâché construction of imagination and theory filled in between a few iron posts of observation." According to this view, all that we can really know is what we observe. All the rest is a creation of the human mind.

This controversy about the nature of theoretical models is analogous to one which has long existed in the field of mathematics. Mathematicians do not agree about the nature of "reality" either, at least not the mathematical kind. Some of them, who are often referred to as Platonists, believe that mathematical knowledge is something that is discovered. They believe that there is a sense in which mathematical objects, such as prime numbers and infinite sets, have a real existence of some kind. They think that it is the task of the mathematician to discover what the relationships between real mathematical objects are.

Other mathematicians think that mathematics is a human invention that is independent of all experience. They feel that mathematicians do nothing more than find logical relationships that connect abstract concepts that human minds have created.

It is not likely that the argument will be settled any time soon. For example, does the number "two" have any real, objective existence? Or is it purely a human invention? No one really knows. The only thing that is clear is that "two" does not exist in the way that an orange or an apple does. We often have the experience of observing two oranges or two bananas, and we can write down the symbol "2." But the number itself is not the symbol. It is an abstract concept that we can never observe, at least not with our eyes. If we perceive it directly, that is something we do only in our minds.

* Naturally, proponents of the superstring theory would disagree.

Similarly, one can regard the concepts of physics either as things that are discovered in reality, or as objects that are invented. One can take the point of view that such objects as quarks, neutrinos, and quantum fields have a real existence. Alternatively, one can believe that they are only things that we have invented to help us make some sense out of the phenomena that are directly observed.

It is probably safe to say that the majority of physicists think of objects like quarks and neutrinos as things which are real in some sense. But they work with these concepts every day of their professional lives. This in itself is sufficient to endow them with a certain kind of reality. However, we should probably not attach much importance to this. Investment brokers experience the same thing when they speak of the stock market as something that can be "jittery" or "confused" or "elated." They know that the market is only an abstraction, yet they endow it with so much reality that they do not hesitate to speak of it as though it could experience human emotions.

If the concepts of physics are human inventions, it does not follow that they can be chosen arbitrarily, or be conceived of as interacting with one another in arbitrary ways. As Wheeler points out, the theoretical constructions of physics must be affixed to "iron posts of observation." Only certain kinds of theoretical constructions are allowable, those that will fit between the posts. One cannot imagine "reality" to be whatever one likes. There are constraints.

If there is a sense in which theoretical concepts are free creations of the human mind, it seems to follow that we will never attain any ultimate truths. As long as we must fill in the gaps between the posts with "theory and imagination," we will never be able to say that we have anything more than an imperfect model. Knowledge of the ultimate laws of nature—if indeed there is such a thing—will continue to elude us.

No one knows which view is more likely to be correct. We cannot say whether mathematics and physics allow us to perceive reality, or whether physical and mathematical "reality" are things that are invented. But let us provisionally accept the latter view and explore it further. It seems to suggest some interesting questions.

To what extent do our mathematics and physics reflect the structure of the human mind? If, at some time in the future, we discovered that other intelligent beings existed in the universe, would we find that their mathematics and physics were similar to ours? Or would we find that, because their minds were different from ours, their scientific conceptions were different too, perhaps so different that we would find them incomprehensible?

It is sometimes assumed that if we were to encounter other intelligent life forms, the first step toward establishing communication with them would be the exchange of mathematical theorems. The idea is that since mathematics is universal, it could be used to create a basis for mutual understanding. However, we cannot be sure that things would work out this way. If mathematical truths lack a universal character, if mathematical concepts are reflections of the ways that we think, such attempts might fail. We might find that because we thought so differently, we could not establish communication with these extraterrestrial beings at all.

Naturally this is not the only possibility. The character of physics could be somewhat different from that of mathematics. It is conceivable that we could find that mathematics did provide a common basis for understanding, but that we and the alien creatures we had encountered saw the physical world through different eyes.

Such imaginary scenarios can be fascinating, and one is consequently tempted to pursue them at great length. However, I am not sure that much would be gained by this. One might end up saying, over and over again, in one way or another, "Either 'reality' is something we observe, or it is, in some sense, a product of our imaginations."

But it might be enlightening to go back to the former point of view, and provisionally adopt *it* for the sake of argument in order to see what some of its implications are. We will now assume, for the sake of argument, that the models created by theoretical physicists are a reflection of reality—not our minds—and that they are becoming so accurate that knowledge of ultimate physical laws is within our grasp.

This suggests a question that has been asked already at several points in this book. Since the answers that have been obtained have not yet seemed to be completely satisfactory, it might be worthwhile to ask the question once again.

If physical models are a reflection of reality, would it be possible to know everything, or at least everything of significance, about physical reality? Or would there still be areas in which knowledge of ultimate truth was still unattainable?

In a previous chapter it was noted that if we are unable to decide whether the standard interpretation of quantum mechanics or the many-worlds interpretation is more reasonable, then we will never know the "truth" about other universes. However, if this is only an isolated case, it may not have all that much significance. And, in any case, we may eventually decide that there are good philosophical reasons for rejecting (or accepting) the many-worlds theory.

Therefore it seems reasonable to try to see if any other "undecidable" questions have cropped up in theoretical physics. If it turns out that they have, we would be forced to conclude that, even if physicists discover—rather than invent—reality, there are parts of reality that they cannot see.

As a matter of fact, examples of such questions are not hard to find. The existence of antiparticles raises an especially intriguing one. It has been shown that an antiparticle is mathematically equivalent to a particle that is traveling backwards in time. If an electron could reverse its time direction, there would be no way to distinguish it from a positron. Thus it is not possible to say whether two kinds of objects—particles and antiparticles—exist, or whether only one kind exists. In the latter case, we would naturally have to assume that quantum particles are free to ignore the distinction between "past" and "future" that governs the behavior of macroscopic objects. However, the equivalence between an antiparticle and a time-reversed particle implies that this may indeed be possible.

Every event that is observed in nature in which antiparticles participate can thus be interpreted in either of two different ways. For example, according to the conventional view, an electron and a positron can annihilate one another. When they do, their mass is transformed into energy, and gamma rays appear in their place.

But there is another way of interpreting the same event. Assume that an electron is traveling forward in time. At some point, it emits a gamma ray, and kicks itself into a time-reversed mode. Its journey into the past as a positron will begin at the same point where the electron's forward progress stopped. Since we, as "time-forward" beings, interpret the particle's behavior from our point of view, we will think that two particles have met one another and experienced annihilation.

In either case, the mathematical description of the process will be the same. The mutual annihilation of an electron and a positron is accurately described by quantum electrodynamics. However, the mathematical description does not tell us which picture is to be preferred. We are forced to admit that it is possible that some of our ideas about the "direction of time" might have to be modified when we deal with subatomic events. There is no reason to think that the description in terms of particles and antiparticles is necessarily preferable.

The concept of "antiparticle" is therefore somewhat ambiguous. Nor is this the only ambiguity encountered in theoretical physics. In fact, there are many of them. For example, it is not exactly clear what is meant when we say that the speed of light is a limiting velocity that only massless objects can attain.

When Einstein published the special theory in 1905, this idea seemed to be a very straightforward one. No massive object could attain light velocity, no matter how strongly it was accelerated. It seemed that there was nothing more to say. But then, during the mid-1960s, physicists Gerald Feinberg and George Sudarshan showed independently that special relativity did not really rule out the possibility that there could be particles that traveled faster than light. If such particles, called *tachyons*, were real, the only restriction imposed by the theory was that they could never slow down to light velocity; they would encounter the same barrier from the other side.

If tachyons exist, they would have to have some very strange properties. For example, if they lost energy, they would move faster, not more slowly. And if their energy was somehow lowered to zero, they would move with infinite speed.

But this does not imply that they cannot exist. As we have seen, many of the particles known to modern physics have strange properties. It is not obvious, after all, that a tachyon is any stranger than a neutrino, which can be viewed as a kind of disembodied spin. Neutrinos, we may recall, are massless, uncharged, spinning particles that move at the velocity of light.

If tachyons exist, then one must assume that they can travel backwards in time. According to the special theory of relativity, all faster-than-light objects must have this property. Thus if beams of tachyons could be employed to transmit signals, it would be possible to send messages into the past.

But this does not necessarily imply that tachyons have no reality. If they were real, but never interacted with ordinary matter, no time-travel or causality paradoxes would be created. It is also possible that they do interact with matter on rare occasions, but that their interactions are random and uncontrollable. Since this would make it impossible to use them to send signals, paradox could be avoided in this case also.

Tachyons have not been detected experimentally, and the hypothesis of their existence does not solve any outstanding problems in theoretical physics. However, although there are not yet any good reasons for believing in their reality, they remain an interesting theoretical possibility.

In my view, the concept of the tachyon is especially intriguing because it is not clear whether special relativity predicts the existence of this hypothetical particle or not. The existence of tachyons is allowed, but it is not required. This raises the following question: Does everything that theory *allows* exist, or should we attribute existence only to that which theory *requires*?

There is, by the way, a rather colorful way of expressing the difference between the two. It has been suggested that nature follows either a "democratic" or a "totalitarian" principle. These principles can be stated in the following way:

Democratic: Everything that is not expressly forbidden is permitted.

Totalitarian: Everything that is not expressly permitted is forbidden.

The two principles are quite different. In many cases, however, they cannot be applied. For example, the existence of virtual particles is required by quantum field theory. If one accepts the theory, it is necessary to conclude that there are virtual particles everywhere. The application of either the democratic or totalitarian principle would lead to the conclusion that they exist. It is only when the principles are applied to entities of an ambiguous nature, such as tachyons, that a distinction can be made.

The questions raised by the hypothesis of the existence of tachyons can be rephrased in yet another way, one that makes these questions seem almost philosophical in nature. One can ask, Does everything that is possible exist, or does reality contain only those elements that are logically necessary? This is the kind of issue that could be debated endlessly. Perhaps it is philosophical in nature. That should not surprise us. After all, there are areas in which science and philosophy border on one another.

Philosophical or not, such questions show that there are ambiguities associated with the concept of "scientific truth." If we are unable to say precisely what kinds of entities exist, then it is not easy to state clearly what physical reality is like or what, exactly, is "true" or "real."

There are yet other kinds of ambiguity that have cropped up in modern physics. The fact that we exist and are able to observe the universe raises some particularly baffling questions. In particular, one can ask, Precisely why is it that the universe is so hospitable to life?

It shouldn't be. The presence of life in the universe seems to depend upon a whole series of improbable coincidences. For example, if life is to be possible, each of the four forces that physicists have observed in nature must have precisely the right strength. If electric forces were only a little stronger than they are, no element heavier than hydrogen could be formed. The positively charged protons would repel one another so strongly that their mutual repulsion could not be overcome by the strong nuclear force. A proton could bind to a neutron, perhaps. However, the formation of a nucleus such as helium, which contains two protons, would be impossible.

On the other hand, if there is to be life, electrical repulsion cannot be too weak. If it were, protons would combine too readily, and nuclear reactions would proceed at a much faster rate than they do. In such a case, stars would not burn slowly and steadily; hydrogen nuclei would combine into helium so readily that stars would explode like thermonuclear bombs. It is hard to imagine that living beings could exist in such a universe.

If the ratio between the strong and weak forces were a little different, the universe would again be inimical to life. Either hydrogen nuclei would combine into helium much too readily, or the reaction would simply not take place. This would produce a universe in which there were no stars, or one in which stars burned out so rapidly that life would not have time to evolve.

It is easy to imagine many other small variations in the laws of physics that would produce universes in which no life could exist. In some of these universes, there would be no atoms. In others, there would be no supernova explosions to spread the elements upon which life depends—such as carbon, oxygen, and nitrogen—through space. Some conceivable universes would contain atoms, but no stars or planets. Others would contain cosmic rays of such intensity that primitive living organisms would not long survive if they did evolve.

The more one considers such possibilities, the more it seems that the existence of life is based on a series of fortuitous accidents. Some of these accidents are not even related to basic physical laws, but rather to the apparently accidental properties of certain nuclei and atoms.

For example, life—at least life as we know it—is based on the element carbon. Carbon is a component of fats, proteins, and carbohydrates. If there were not a great deal of it on the earth, we would not exist. Yet the existence of abundant quantities of carbon in the universe seems to be the result of pure accident.

The carbon that exists today was originally synthesized in nuclear reactions that took place in the interiors of massive stars that were later to explode as supernovae. Carbon was synthesized as follows: First, two helium nuclei combined to produce a nucleus of the element beryllium. If the beryllium managed to capture a third helium nucleus before breaking apart, a carbon nucleus would then be created. However, the isotope of beryllium that is created when two helium nuclei combine is very unstable; it splits up into helium again in about 10^{-16} seconds. Thus, if carbon is to be created, the beryllium nucleus must capture the third helium nucleus quickly. If it does, a stable isotope of carbon will be formed.

One would not expect such a process to take place very often. It seems just too improbable. However, seemingly by accident, the helium-beryllium system has an energy level that just "happens" to have the value needed to speed this reaction along. If this energy level were just a bit different, only minute quantities of carbon would exist.

Perhaps we have no right to assume that life must necessarily be carbon-based. It is possible to imagine beings whose bodies depend upon a silicon-based chemistry, for example. There are reasons for doubting that such beings exist. Silicon is a heavier element than carbon, and it is possible to argue that an evolution of life that was based on it would be more difficult or less natural. But perhaps this view only reflects our carbon "chauvinism." It is possible that beings made of silicon could be skeptical of the idea that carbon might be a building block for life.

However, if one looks into the matter in detail, one finds that the existence of large quantities of silicon depends upon "accidents" too. In fact, the most overwhelmingly probable kind of universe would be one in which no elements heavier than hydrogen and helium existed, or one in which they were present only in the cores of planetless stars.

The idea that the existence of life in the universe is a fact that needs to be explained is called the *anthropic principle*. "Anthropic" means "relating to mankind." Naturally, it is the existence of life, not specifically human life, that is a puzzle. However, since it is human scientists who wonder about the matter, the term may not be entirely inappropriate.

Some philosophers have criticized the use of the principle on the grounds that this represents a return to a pre-Copernican outlook. Since Copernicus showed that the earth was not the center of the cosmos, it has been considered unreasonable to conclude that the universe appears to have been designed specifically for man. Yet the anthropic principle seems to lead inescapably to this conclusion, or to one very much like it. Perhaps the principle does not imply that the universe was made for human beings, but it does seem to lead us to the conclusion that something must have caused it to be constructed in such a way that it would be hospitable to life.

The anthropic principle may indeed have a pre-Copernican

flavor. However, there does not seem to be any obvious way to evade the principle's implications. We must deal with the fact that if the laws of probability have any meaning, the odds are overwhelmingly in favor of a universe that is devoid of life. Since the universe very obviously is not lifeless, this is something that must be explained.

If one accepts the idea that the principle points out something of significance, two very different kinds of conclusion are possible. One can conclude either that life exists out of logical necessity or, alternatively, that there are an infinite number of universes. In the latter case, the presence of life in our universe would not be surprising. The lifeless universes would presumably exist in great numbers, but there would be no one to see them.

At first glance, this conclusion seems a fairly reasonable one. After all, it is possible that there are numerous other universes. Even if the many-worlds interpretation of quantum mechanics is incorrect, these universes could have been created by quantum fluctuations of the sort that presumably led to the creation of our own.

On the other hand, recent discoveries have made the alternative, the idea that the conditions necessary for the evolution of life are somehow written into the laws of nature, seem less absurd than it did a few years ago. For example, when the anthropic principle was first proposed, it was possible to argue that if life was to be created, then the expansion rate of the universe had to be very finely tuned.

It was possible to consider the state of the universe when it was just a few seconds old, and to consider what would have happened if the expansion had been just a little slower, or a little faster. When this was done, it appeared that one could construct arguments about the improbability of life that were similar to some of those that I have given.

If the expansion rate at this time had been greater than it was by one part in a million, then stars and galaxies would never have formed. The matter in the universe would have been propelled outward with such velocity that gravity would never have had a chance to condense it into the clumps of matter that were later to become stars and galaxies. In such a case, the universe would contain nothing but rarefied gas.

On the other hand, if the speed of the expansion had been one part in a million less, the universe would no longer exist. The expansion would have come to a halt when it was only about 30,000 years old. The matter in the universe would not have dispersed fast enough to escape the retarding effects of gravity, and a state of contraction would have set in before it had cooled to a temperature below 10,000 degrees. Since it is hard to imagine that life could have evolved in so short a time, under such extreme conditions, one must conclude that such a universe would be as lifeless as one that expanded too rapidly.

However, it is no longer possible to conclude that the fine-tuning of the expansion rate was an accident. According to the inflationary universe theory, the expansion had to proceed at just the right rate to produce the kind of universe that we observe today. The fact that the expansion of the universe was neither too great nor too small is, according to the theory, a consequence of the fact that the quantum fields which permeated the early universe brought about a period of inflationary expansion.

If the fine-tuning of the expansion rate was no accident, the argument about the improbability of life loses its force. One cannot say that a universe with the right expansion rate is improbable if that is the only kind of universe that can exist.

It is possible that we will eventually discover that the fine-tuning of the strengths of the forces is no accident either. A theory which unified all four of them would presumably tell us why each force has a particular strength. In such a case, we would have to conclude that the conditions necessary for the evolution of life were an automatic consequence of the laws of physics.

The implications of this are staggering. It is not easy to say what one would make of the fact that the existence of conditions for the evolution of life were written into the laws of physics. And what if it turned out that only one kind of unified-force law was logically possible? In such a case, we could paraphrase Einstein and say that life exists because God had no choice as to whether or not He would create it.

At one time, philosophers and theologians often argued that the evidence of design in the construction of the natural world implied the existence of a Creator. Such arguments are not very fashionable nowadays. They have not been ever since Kant pointed out that this *argument from design* could as easily be used to argue for the existence of a demiurge. However, if ours is the only universe that exists, one has to explain the nature of physical law some way, and it seems natural to appeal to the idea of a Creator.

On the other hand, the idea of a Creator would become superfluous in the case that only one kind of physical law is possible. If God had no choice as to whether or not life would be created, the hypothesis of His existence is unnecessary.

One can avoid confrontation with such metaphysical and theological enigmas if one inclines to the opinion that many kinds of universe are possible. If the laws of physics vary from one to another, such problems do not arise. In such a case, the laws of nature would have nothing to do with logical necessity. They would have the form we observe purely by chance.

There is no evidence which would indicate that the hypothesis of many universes is correct. I suspect that one's philosophical preconceptions will cause one either to accept it or reject it. For example, a theist would probably prefer the hypothesis that there is one universe, and that it is governed by physical laws of divine origin. An atheist, on the other hand, might prefer the many-universes theory. Finally, an adherent of some such eastern religion as Taoism might prefer the idea that natural laws are as they are out of logical necessity, and perhaps even identify this logical necessity with the Tao itself.

But perhaps it would be best if I did not pursue this line of thought too far. If I did, we might quickly become embroiled in a lot of murky theological arguments and lose sight of the fact that the topic under discussion is supposed to be the nature of scientific truth. In any case, there is no one-to-one correspondence between one's religious beliefs and the kind of cosmos that one must accept. One could be a theist, for example, and still accept the many-universes theory. After all, if God could create one universe, He might have created many.

This chapter began with a question about the nature of scientific truth. It appears that pursuing this question has led us down a crooked path, and has not produced much in the way of conclusions. Asking whether exact laws of nature can be discovered has led to other questions that seem to have more of a philosophical than a scientific character.

One should not feel surprised that we have encountered philosophical questions about the nature of theoretical models, or about the existence of ambiguities in physical theory. Scientists have often become preoccupied with questions about the meaning of scientific discoveries.

It is somewhat more surprising to find physicists discussing such questions as the implications of the existence of life. The fact that there are metaphysical implications associated with this topic is even more astonishing.

But perhaps it is no accident that we became embroiled in philosophical and metaphysical ideas. In my opinion, physics and cosmology have made such strides in recent years that they have found themselves encountering questions that were previously thought to be purely philosophical in nature. If one asks enough questions about the nature of reality, and tries to obtain a clear picture of all the facets of the natural world, one eventually finds oneself asking what reality "really is." And of course, this is a question that has always been considered to be in the domain of philosophy.

But perhaps we should remember that all scientific questions were once considered philosophical. Plato discussed questions about the natural order in his dialogue *Timaeus*. No one thought this an odd activity for a philosopher to engage in. Not many years later, Aristotle philosophized about the nature of the heavens, and about many other kinds of natural phenomena as well. Even in Newton's day, there were no "scientists," only "natural philosophers." In fact, the title of Newton's most important book, the one in which he put forward his laws of motion and his theory of gravitation, was *Philosophiae Naturalis Principia Mathematica* (*Mathematical Principles of Natural Philosophy*).

Since the time of the classical Greeks, many questions that were once considered to be philosophical in nature have become scientific ones. The reason is obvious. Many matters about which one could once only speculate are now amenable to empirical investigation. As knowledge has increased over the centuries, science has taken more and more topics away from philosophy.

But perhaps, as science seeks to extend the frontiers of knowledge, it may find itself completing a circle, and returning to its origin. About two and a half thousand years ago, the Greek philosophers began to ask questions about the nature of the universe. As time passed, science was able to provide tentative answers to some of these questions. As knowledge increased, some of these tentative answers became firm ones; others were discarded. Now, finally, we seem to have reached the point where science knows so much, and has answered so many questions, that it has confronted questions about the nature of being itself. And of course, questions about the nature of being were the very ones that were asked by the classical Greeks.

9

The Nature of Reality

When Isaac Newton discovered the inverse-square law of gravitational attraction, he found a theory which explained numerous different kinds of phenomena. His gravitational law could be applied to the motion of projectiles near the surface of the earth, to the motions of celestial bodies, and to the ocean tides. Yet neither Newton nor his contemporaries were completely satisfied. Although Newton's law could be used to make accurate numerical predictions that were confirmed by observations, it did not explain precisely how gravitating bodies attracted one another. The best that could be done was to speak of "action at a distance." But no one could explain just why this action at a distance should exist.

After Maxwell had propounded his electromagnetic theory in 1873, scientists had a theory which explained all known electric and magnetic phenomena. Furthermore, the theory showed that light could be explained as electromagnetic waves. Like Newton's, Maxwell's theory yielded predictions that could be confirmed by experiment. But again, physicists were not satisfied. They demanded something more. They wanted to know how electromagnetic waves propagated through space.

Since they could not conceive of waves which traveled through a vacuum, they appealed to the old idea of an ether that supposedly permeated all space. The hypothesis of the existence of an ether did not add much to Maxwell's theory. The theory's predictions were the same whether one invoked the presence of this imaginary substance or not. And yet, even when the ether hypothesis began to produce contradictions, physicists were reluctant to give it up. It took an Einstein to demonstrate that the concept was superfluous.

Since the development of quantum field theory has been discussed in previous chapters, there is no need to explain again how the problems of action at a distance and the propagation of light were eventually solved. The question I want to discuss here is a somewhat different one, one that bears on the nature of theories and of the reality that they attempt to describe.

Newton's theory of gravitation and Maxwell's electromagnetic theory are among the most successful in physics. Although we know today that they are not perfectly accurate, they are excellent approximations that can still be used whenever relativistic or quantum effects are unimportant. Both theories must be considered significant advances, and Newton and Maxwell must be counted among the greatest physicists of all time.

If the theories were so successful, why was it felt that something was lacking? Why did Newton's critics attack the concept of action at a distance so vehemently? Why did Maxwell and his contemporaries feel that the superfluous idea of the ether had to be introduced? Isn't it enough to create a theory that will allow one to predict experimental results accurately? What more can one ask?

Before I attempt to answer these questions, it might be illuminating to look at yet another example, one that deals with a theory developed by one of the cultures of the ancient world. Comparing this theory with modern ones will clarify some of the issues involved and throw some light on the question, How much does a good theory have to explain?

The ancient Babylonians were excellent astronomers. They

were able to predict planetary movements and the occurrence of eclipses with near-perfect accuracy. Although they lacked modern astronomical instruments, such as the telescope, the precision of their observations was not exceeded until the late nineteenth century. The Babylonians made up for their relative lack of technology by making observations over a period of centuries. By observing the stars continuously, year after year, decade after decade, they were able to obtain results that remained unsurpassed for more than two millenia.

The classical Greeks, on the other hand, accomplished little in the field of astronomical observation. For the most part, they depended upon compilations of Babylonian data. Yet Greek astronomical theory seems to us to be "scientific," while the Babylonian does not.

The reason is not hard to discern. The Babylonians amassed large quantities of data, and successfully used this data to make predictions. But they stopped there. They did not know why celestial objects moved the way that they did. Perhaps they did not want to ask, since they regarded the heavens as divine. The Greeks, on the other hand, went about things differently. They attempted to see beyond observed phenomena, and to propound theories to explain the celestial motions.

Most of the Greek astronomical theories were wrong. The Greek philosophers and mathematicians generally rejected the hypothesis, proposed by the followers of Pythagoras in the fifth century B.C. and by Aristarchus of Samos in the third century B.C., that the earth revolved around the sun. Instead, they thought up various kinds of geocentric schemes. According to Plato's pupil Eudoxus, for example, the heavens were made up of a system of moving crystalline spheres which carried the planets in their orbits around the earth.

To modern eyes, such an idea seems implausible in the extreme. However, we refrain from criticizing Eudoxus and those who elaborated upon his system too harshly. We see the Greeks as scientists who were attempting to understand the physical events which produced observed phenomena. We admire them for attempting to penetrate beyond appearances and to gain an understanding of the nature of physical reality.

Contemporary scientists are motivated in the same way that

Eudoxus was. It is surely not going out on a limb to say that the reason most people become scientists is that they experience intellectual curiosity about the physical world. This curiosity exhibits itself in many ways; it is one of the things that is responsible for the fact that, even after a theory has been confirmed by experiment, scientists will often continue to engage in debate on the subject of what the theory really means. In other words, they still want to know what kind of picture of reality it implies. This is the reason that they argued about Newton's action at a distance even when it was obvious that the inverse-square law worked. This is why they asked precisely how electromagnetic fields were propagated through space.

Sometimes the problem of creating an accurate picture of reality is difficult indeed. The scientists of Newton's time never found out what gravitation was. They were baffled by action at a distance. Those of Maxwell's day thought that the concept of the ether would provide a solution to their problem. And of course they were proved wrong in the end.

Problems of a similar nature exist today. Even after a theory has been confirmed by experiment, the task of interpreting it can be difficult. For example, it would not be inaccurate to say that no one really understands the nature of the electron, even though physicists experience little difficulty predicting how electrons will behave.

According to quantum electrodynamics, it is necessary to consider an electron to be an object of infinite charge and infinite mass that is screened by a cloud of virtual particles. When QED was first proposed, few physicists took it seriously. They didn't believe that a theory which invoked the existence of such infinite quantities could possibly be an accurate representation of reality. However, when it was shown that QED could be experimentally confirmed to a degree of accuracy that was practically unheard of in other fields of physics, they had to take it seriously.

Possibly, the infinities that one encounters in QED are really no barrier to understanding. After all, it is a bare electron, one that has been stripped of its virtual particles, that has the infinite charge and infinite mass. But bare electrons do not exist in nature. All of the electrons that one can observe are screened. So perhaps we do not really have to worry about what the "real" charge and mass of an electron are. The only quantities that one has to deal with in practice are the charge and mass of the electron *and* its virtual particle cloud.

What is an electron "really" like? There seem to be a number of possible answers to this question. It may be that the concept of a bare electron is meaningless, that it is not valid to think of a particle without taking its interactions with fields into account. Perhaps the renormalization procedure, as mathematically suspect as it seems, is telling us something about nature, that combinations of particles and virtual particles are the only things that are physically real.

On the other hand, perhaps the infinities that plague QED would disappear in a better theory. At the moment it is hard to say what such a theory would be like, or how the infinities might be eliminated. Perhaps we will discover that the electron is not truly elementary, that, like the nucleon, it is a composite particle. At the moment, there is not the slightest bit of evidence for this conjecture, but that does not mean that it is impossible.

Alternatively, the infinities of QED might have something to do with the unknown quantum events which presumably take place in the Planck region. QED conceives of the bare electron as a structureless particle whose charge and mass are concentrated in a mathematical point. The electron is viewed as a particle that has no physical dimensions. There is really no way of knowing how accurate an approximation this is. However, if the physical dimensions of an electron are less than the Planck distance, we may not know what the particle "really" is until we gain an understanding of the quantum fluctuations that might affect the nature of space and time.

The lack of understanding of the "real" nature of the electron and of the origin of the infinite quantities that are encountered in QED are not viewed as major problems, however. Quantum electrodynamics and the quantum field theories that have been modeled upon it, such as QCD, have been so successful that physicists have felt justified in putting these questions aside. Quantum field theory has explained so much about the fundamental nature of reality that it is possible to say that the questions that have been answered are much more numerous than those which have been temporarily swept under the rug. Considerable progress has been made toward the construction of a unified theory of the forces. The existence of new particles has been predicted, and these particles have been observed. Finally, the new theories have dramatically increased our understanding of the nature of matter and of the processes which took place in the early universe. One must conclude that, in spite of a few lingering problems, the new quantum physics has been enormously successful.

The other theoretical ambiguities that have been encountered are not a source of great concern either. For example, it is not really necessary to worry overmuch about the possible existence of tachyons in the absence of any evidence that they are real. The fact that they are allowed by special relativity does raise some intriguing questions. In particular, there is the question of the extent to which we can ascribe reality to something that we may never be able to observe. However, the existence or nonexistence of tachyons has little or no bearing on our understanding of the particles of ordinary matter.

The discovery that an antiparticle is mathematically equivalent to a particle that is moving backwards in time has a bit more significance. Unlike tachyons, antiparticles are observed in the laboratory every day. It is a bit disconcerting to discover that there is no way of telling whether there are two kinds of objects—particles and antiparticles—in the universe, or whether there are only particles that can travel in either time direction.

Or perhaps it is a mistake to make a distinction between the two viewpoints. Perhaps they are really the same. It is conceivable that this result is telling us something about the nature of time that we have not yet grasped. Time, after all, is one of those elusive things that we all think we understand, but which may contain mysteries that we have not yet fathomed.

In any case, this is also a problem that can be put aside. Although a certain amount of ambiguity exists, it is possible to say that the behavior of antiparticles is reasonably well understood. Quantum field theory tells us how particles and antiparticles are created and destroyed. And if the predictions of the grand unified theories are eventually verified experimentally, we may even be able to say that we understand why there are more particles than antiparticles in the universe.

On the whole, the picture of reality that modern physics gives us must be said to be reasonably satisfying. If there are a few unanswered questions, and a few problems of interpretation, that is only to be expected. In fact, we should probably be happy that they exist. The existence of unsolved problems often suggests lines of research that eventually lead to the discovery of new and better theories. If everything appeared to be understood, there would be little for scientists to do.

It is necessary to distinguish between problems of interpretation and the problems that arise when there is a discrepancy between theoretical predictions and experimental results. However, both kinds of problem can suggest new lines of research that lead to new discoveries. A scientist who finds himself wondering whether the picture of reality that is implied by a particular theory can really be reasonable is just as likely, perhaps even more likely, to hit upon a revolutionary new idea as one who tries to understand why an experiment has not turned out the way theory says it should have.

The problems that I have discussed so far in this chapter are more or less typical of those which have arisen throughout the history of physics. The scientists of the seventeenth century did not know what to make of action at a distance. We do not know how to interpret the fact that an antiparticle moving forward in time and a particle moving backwards are mathematically equivalent. Physicists of the late nineteenth and early twentieth centuries found that the ether they thought so necessary had to have properties that were quite paradoxical. When we try to use quantum electrodynamics to determine the bare mass and bare charge of an electron, we obtain a paradoxical result. Michelson and Morley could not detect the effects that the motion of the earth should have had on the propagation of light. We have not yet detected the proton decays that the grand unified theories say should take place.

However, there is one problem faced by physicists today that is not typical. Perhaps the difference is only one of degree. Nevertheless, this problem is one that is so fundamental that it cannot be put aside as, for example, questions about the properties of the electron can. It is a problem about which physicists have been arguing since at least the 1920s and 30s. Bohr, Einstein, Heisenberg, Dirac, Schrödinger, and many other notable physicists have all wrestled with it, and yet we are no nearer a solution than we were when they started.

The problem is this: No one really understands the meaning of quantum mechanics.

Perhaps it would not be a bad idea to emphasize, again, that quantum mechanics is an amazingly successful theory. Its predictions have been experimentally confirmed over and over again. Furthermore, there are numerous other theories that are based on quantum mechanics that have been experimentally confirmed also. Although quantum mechanics may be elaborated upon still further in the years ahead, it is inconceivable that it will suddenly be overthrown. Naturally no one knows what the next century or succeeding centuries will bring. However, it would not be an exaggeration to say that it is no more likely that quantum mechanics will be replaced by some new theoretical conception in the immediate future than it is that historians will suddenly discover that George Washington was really a woman. If quantum mechanics does eventually give way to some new kind of theory, the chances are that physicists will continue to use it under most circumstances, just as they continue to use Newtonian mechanics when they do not have to deal with objects that move at relativistic velocities.

The problem is that quantum mechanics can be interpreted in a number of different ways, and that no one is sure which interpretation is most likely to be correct. Nor is there any general agreement about the theory's implications. Quantum mechanics can be used to construct several different pictures of reality. But all of these pictures contain elements that appear to be a bit paradoxical. It appears that whichever interpretation one chooses, the implications of quantum mechanics are not easy to accept.

The problems associated with interpreting quantum mechanics stem from the fact that the properties that subatomic particles exhibit seem to depend upon the kind of experiment that one chooses to perform. For example, all elementary particles have wave characteristics. It is possible to perform experiments in which the wave character of the particles is exhibited. It is also possible to do experiments in which they are observed as discrete particles. But it is not possible to observe them as waves and as particles at the same time.

According to the interpretation developed by Max Born, wave amplitudes are related to the probability that a particle will be found at a particular position in space. But we should not be deceived into thinking that the wave formalism is only a kind of mathematical shorthand for calculating where a particle will be found. Quantum waves must be regarded as physically real entities.

During the mid-1920s, a phenomenon called *electron diffraction* was discovered by George Thomson and by the American physicists Clinton Davisson and Lester Germer. Thomson and the team of American experimenters independently observed that interference patterns could be obtained when beams of electrons were made to bounce off a crystal onto a photographic plate. When the image formed by the electrons was examined, it was found to be made up of light and dark bands that were very similar to those which were formed when a crystal was bombarded with X rays.

In the case of X rays, the pattern of light and dark is easily explained. X rays are electromagnetic waves, and waves can interfere with one another. If two waves are superimposed so that crests match up with crests and troughs with troughs, the result is a single wave of twice the amplitude. If, on the other hand, crests match up with troughs, the two waves will cancel one another out, and "darkness" will result. Since waves that bounce off the atoms of a crystal in certain directions tend to reinforce one another, while waves bouncing in other directions cancel, a pattern of light and dark bands is the result.

Wave interference is a common phenomenon, one that can be seen with no scientific apparatus whatsoever. Anyone who wants to observe the interference of light waves can do so by looking at a light source (a light bulb, for example) between two fingers that are held slightly apart. If the fingers are then gradually brought closer together, light and dark fringes will appear just before they touch.*

---

* Readers of my previous books may note that I have used this example before. I admit that I must plead guilty to the charge of plagiarizing myself. However, I think that the offense is mitigated by the fact that there are not many experiments in physics that can be performed with no equipment whatsoever. This makes the finger experiment almost unique, and worth mentioning more than once.

On the other hand, there are numerous experiments in which electrons appear as particles. When an electron passes through a bubble chamber, for example, it leaves a track of a sort that can only be produced by particles. When an electron strikes a fluorescent screen, a tiny flash of light will be produced. Such behavior is not characteristic of waves. A wave could not be localized in one place; it would spread out over the screen.

If it were possible to say that electrons were neither waves nor particles, but had characteristics of both, matters would be much simpler. We could invent some new term, "wavicle," perhaps, to describe them. Unfortunately, an electron does not have this character. Depending upon the experimental arrangement that we use to observe it, the electron always manifests itself as one or the other. We never see it as a wave-particle hybrid.

The puzzling thing about this is that it isn't the electron that "decides" whether it will appear as a wave or as a particle, but the observer. It is the experimenter who must choose—by setting up one kind of apparatus or another—what characteristics the electron shall have.

The wave-particle duality is not the only one that can be observed. For example, the uncertainty principle implies that the observer can choose to measure the momentum of a particle or its position, but never both simultaneously. Again, the particle's characteristics seem to depend upon the manner of observation.

These two kinds of duality are related. When we observe an electron as a particle, we find that it has a definite location in space at a given moment of time. For example, when the electron strikes the fluorescent screen, the flash of light has a definite location. Waves, on the other hand, possess momentum, but they have no precise location in space. A wave is a series of crests and troughs; it must always be spread out over some finite volume.

The interaction between a quantum particle (or wave) and an observer becomes even more mysterious if we attempt to create mental pictures of the events that take place on a subatomic level. For example, suppose we have an electron wave. It is a characteristic of waves that they not only occupy a finite volume, but that this volume also increases with time. In other words, waves propagate. If one disturbs the water in one end of a bathtub, the water throughout the tub will soon be in motion. Similarly, an electron wave will spread, so that at every moment of time, the particle's position will be a little more uncertain than it was at the previous moment.

But suppose that we now decide to observe the electron as a particle. The only way to describe what happens to the electron waves when the observation takes place is to say that they suddenly collapse into a particle configuration. The electron no longer has the extension in space that it possessed a moment before. Waves of probability have collapsed into certainty.

There are many other examples of this kind of behavior. All seem equally mystifying. Suppose, for example, that an electron collides with an atom. If we use the wave picture, we must assume that after the collision takes place, the electron wave is scattered in all directions. Obviously, the electron must bounce off the atom. But the direction in which it bounces is a matter of probability. Thus part of the electron wave will spread out to the left, and part to the right; part of it will continue on in the electron's original direction, and part will converge backwards. In fact, after the collision has taken place, the electron wave will be something like an expanding sphere.

If the waves from a number of such collisions are allowed to interfere with one another, an interference pattern results. But suppose that an experimenter does not wish to observe an interference pattern. If he does not, he can perform an experiment that detects the electrons as particles; for example, by allowing them to form tracks in a bubble chamber. If he does this, he will find that each electron is found to be traveling in one direction, not in many; a wave pattern that spreads out in all directions in space has been converted into a particle that moves in a specific direction. Naturally it is impossible to predict what this direction will be beforehand; the indeterministic nature of quantum mechanics makes this impossible. The best that one can do is to say that if many electrons are thus observed, a certain number will be seen to be moving to the right, a certain number to the left, and so on.

Such descriptions of electron behavior sound contradictory. Indeed, the wave and particle descriptions are incompatible. Waves and particles are two entirely different kinds of phenomena. It is not difficult to see why many physicists laughed when

Einstein suggested, in 1905, that light had wave and particle characteristics simultaneously, and laughed again when de Broglie introduced his electron wave theory twenty years later.

But, however outrageous an idea may seem, we must accept it if it is forced upon us by experiment. If experiments demonstrate conclusively that both wave and particle characteristics exist, the only thing that we can do is to accept this fact. If they show also that the existence of either wave or particle character depends upon the type of experiment that we choose to perform, we must try to make sense of this strange result as best we can.

A number of different interpretations of quantum mechanics are possible. However, there is one that is more widely accepted than any other. This is called the *Copenhagen interpretation,* because it was developed by Niels Bohr and his colleagues in discussions that took place at Bohr's Institute for Theoretical Physics in Copenhagen.

It is often said that the Copenhagen interpretation is the one that is accepted by the majority of physicists. I suspect that it would probably be best to modify this statement somewhat and to say that it is adhered to by the majority of the physicists who have pondered the problem of the correct interpretation of quantum mechanics. I often suspect that most physicists are agnostics, and that they don't worry much about interpretations at all. It is not necessary to interpret quantum mechanics in order to make use of the theory. It is perfectly possible to carry out calculations without ever wondering what it means. One no more needs to know what kind of picture of reality is implied by quantum mechanics than Newton needed to know what action at a distance was before he could compute planetary orbits.

But this does not imply that the problem of interpretation is trivial. It is not. Bohr once commented that "those who are not shocked when they first come across quantum mechanics cannot possibly have understood it." This statement is as true today as it was when he uttered it. However one interprets the theory, one obtains a picture of reality that is odd indeed.

The Copenhagen interpretation is based on two tenets. The first is the principle of complementarity, which Bohr first enunciated in a lecture that he delivered in Como, Italy, in 1927. According to Bohr, it was possible to describe quantum reality only in terms of mutually exclusive concepts. The wave and particle descriptions, for example, were said to be complementary to one another.

Bohr warned that any attempt to discard the wave and particle pictures and to replace them with something new was misguided. Both had to be retained, because the only contact the physicist had with the quantum world was through his experimental apparatus. Since this apparatus was part of the macroscopic world, the results of quantum experiments could be expressed only in terms of concepts that had been developed to describe that world, such as "wave" and "particle."

To Bohr, complementarity was more a point of view than it was a precisely defined physical law. At different times, he spoke of the principle in different ways. However, it would not be inaccurate to characterize his principle as an insistence upon the idea that quantum reality could only be understood as a fusion of apparently irreconcilable quantities. It is perhaps not irrelevant that, although Bohr was no mystic, he was so intrigued by the Chinese yin-yang symbol that he included it in the coat of arms that he designed when he was knighted in 1947. He apparently felt that the Chinese idea of interpenetrating opposites was a parallel of his ideas about quantum mechanics.*

The second tenet of the Copenhagen interpretation can be stated as follows: One cannot meaningfully assign specific properties to quantum objects without specifying an experimental arrangement by which these properties can be measured. If one wants to speak of "momentum" or "position" or "wavelength," it is necessary to state precisely how those quantities are to be determined.

The Copenhagen interpretation thus implies that quantum particles do not possess the kind of objective reality that we habitually attribute to macroscopic bodies. An electron cannot be said to have a position or a momentum until one of these quantities is measured. Furthermore, it is not possible to say what a particle is doing when one is not looking at it. For example, if an electron is propelled toward a fluorescent screen, it may be possible to de-

---

* One should not read too much into this. Bohr never suggested that there was any real parallel between physics and Eastern mystical traditions.

scribe the device that emitted the electron, and it may be possible to speak of the events that take place when the electron strikes the screen. But one cannot speak of the trajectory that the electron followed. An unobserved electron lacks tangible reality.

Bohr concerned himself with the problems of interpreting quantum mechanics all his life. He expressed himself on the subject in a number of different ways. One of his most radical statements on the topic was a remark he once made when his assisstant Aage Petersen asked him if he thought that quantum mechanics mirrored some underlying reality. Bohr replied:

> There is no quantum world. There is only an abstract quantum physical description. It is wrong to think that the task of physics is to find out how nature is. Physics concerns what we can say about nature.

Most physicists would not go quite this far. Although they would agree with Bohr's contention than an electron does not possess an objectively real momentum or an objectively real position except when these quantities are measured in an experiment, they point out that particles do have certain invariant properties that are independent of observation. Two examples would be charge and mass. Though we don't really know what the bare charge and bare mass of an electron are, or even if those terms are very meaningful, the observed charge and mass of an electron are always the same when the particle is observed from a distance.* As Max Born pointed out, "Though an electron does not behave like a grain of sand in every respect, it has enough invariant properties to be regarded as just as real."

According to the Copenhagen interpretation, the indeterminism of quantum mechanics and the nonobjectivity of such properties as position and momentum are related. It is lack of knowledge about these properties that prevents us from knowing what the particle will do in the future.

According to Newtonian mechanics, it should be possible to predict the future behavior of an object if one knows enough

---

* At very near distances, the cloud of virtual particles is partially penetrated and charge and mass appear to change.

about its position, velocity and direction of motion. For example, there are never any problems associated with calculating the orbit of a planet or the trajectory of a space vehicle if all these quantities are known.

When one is dealing with a quantum system, such prediction is not possible. One cannot say what the position and velocity of an unobserved electron are. If one attempts to measure these properties, they cannot be determined simultaneously. Since momentum is the product of mass and velocity, uncertainty in momentum implies that the velocity is uncertain too. Thus one can never predict what an electron's future trajectory will be. It is only possible to describe it in terms of probabilities.

The nonobjectivity of certain properties of quantum particles has important implications for our outlook on reality. Indeed, the development of quantum mechanics has made certain changes in this outlook necessary. For example, during the early decades of the twentieth century, it was customary to think of such microphysical objects as electrons and atoms as the only things that were "really real." The British astronomer Arthur Eddington pointed out, in one of his books, that a table was not what it seemed. The naive observer might see it as a solid, unmoving object. In reality, Eddington said, the table was mostly empty space. It was really composed of widely separated protons and neutrons around which the electrons revolved. Not only were the electrons in rapid motion, the atoms themselves vibrated rapidly back and forth whenever an object contained any amount of heat.

But if one accepts the Copenhagen outlook, Eddington's interpretation must be discarded. According to the Copenhagen outlook, electrons and atoms are less real than a table. It is all very well to say that a table is composed of an ensemble of subatomic particles. But if we do, we find that we cannot say where the particles are, how fast they are moving, or how they will behave in the future. It may be that the naive view of a table as a solid, objectively real object is the most reasonable one after all.

The indeterminism of quantum mechanics and the idea that properties such as position and momentum are not objectively real quantities have not been accepted by all physicists. Many of them have felt uncomfortable with such ideas, and have sought alternatives to the Copenhagen interpretation.

The most illustrious opponent of the Copenhagen outlook was Einstein. Over a period of years, he engaged in intermittent debate with Bohr, attempting to find ways to show that quantum indeterminacy was an illusion. For example, he invented complicated arguments in an attempt to show that it was illogical to assume that position and momentum could not be determined simultaneously.

Einstein presented one argument after another, generally in the form of imaginary thought experiments. But every time that he did, Bohr was able to find a hidden fallacy in his reasoning. Eventually, Einstein had to admit that his attempts to prove the inconsistency of the Copenhagen interpretation had been unsuccessful. But he still would not give up. He continued to insist that the indeterminism of quantum mechanics implied that it could not be a complete theory. He continued to hope that a better theory could be found that would cause the indeterminism to be discarded.

No such theory was discovered. Einstein still persisted. In his later years, he adopted the point of view that quantum mechanics did not describe the behavior of individual particles. In his opinion, it described only the statistical behavior of ensembles of particles or ensembles of systems. However, many contemporary physicists question Einstein's position. They say that it is not possible to distinguish between a theory which predicts statistical averages and one that gives the probabilities of individual events. In their opinion, finding the probability that a particle will behave in a certain manner and predicting the average behavior of a group of such particles are logically equivalent.

It is sometimes said that Einstein was the most profound advocate of the concept of hidden variables. There is some truth in this statement, even though Einstein never tried to construct, and never advocated, any particular hidden-variable theory.

As we have seen, hidden variables are quantities that could be used to predict the future behavior of a quantum object or system. For example, if an electron possessed a real position and a real momentum at any moment, position and momentum would be hidden variables. They would be "hidden" because quantum mechanics does not allow us to determine what they are.

When Einstein argued against the uncertainty principle, or insisted that the indeterminacy of quantum mechanics would disappear in a more complete theory, he was saying in effect that he believed that the construction of some hidden-variable theory was possible.

We now know that Einstein was wrong. The construction of a valid hidden-variable theory is not possible, at least not if one wants to retain the idea of locality. Bell's theorem and the experiments that have been performed to test Bell's inequality rule this possibility out.

This leads us to the conclusion that it would be possible to construct a hidden-variable theory only if locality were abandoned. Hidden variables could exist if quantum objects were able to exert influences on one another that traveled faster than the velocity of light. If these influences operated in such a way that it was impossible to use them to send faster-than-light signals in the macroscopic world, no paradoxes would result. However, no one really knows what such influences would be like, or how they would operate. However, they do provide an interesting loophole, and it is not possible to assert that the idea of hidden variables can be completely ruled out.

There seem to be fewer advocates of the hidden-variable interpretation of quantum mechanics than there were perhaps around 1970. On the other hand, another alternative to the Copenhagen outlook, the many-worlds interpretation of quantum mechanics, seems to have been gaining ground. Since the many-worlds interpretation has already been discussed, it would be superfluous to go over the same ground again. However, it might not be a bad idea to try to see just what the connection between the Copenhagen and the many-worlds interpretations is. After all, when one first encounters the latter, it is hard to imagine that it could be an alternative interpretation of the same experimental and theoretical results that the Copenhagen outlook is designed to explain.

One way of illustrating the relationship between the two would be to go back to an example that was used previously, that of the electron that is scattered by an atom. As we saw, the probabilistic character of quantum mechanics ensures that if we observe the electron as a particle after the encounter has taken place, it may be found to be moving in any direction. The act of observation gives reality to one possibility out of many.

The various directions in which the electron may travel away from the atom correspond to a number of different possible worlds. One can say that the quantum probabilities describe a number of different potential results. When an observation is made, one of these potentialities becomes real.

According to the many-worlds interpretation, on the other hand, all of these potentialities are realized. There are not a number of different possible worlds, but an ensemble of real ones. In the many-worlds interpretation, the electron does have objectively real properties. The act of observation does no more than tell us which of these alternate universes we inhabit.

The advocates of the many-worlds theory maintain that it is to be preferred to the Copenhagen interpretation because it is logically simpler. They say that there are profound difficulties associated with the Copenhagen outlook. In particular, they point to the problems of describing just what takes place when an observation causes one of a number of different possibilities to become real.

Indeed, the advocates of the Copenhagen interpretation have been unable to agree as to precisely how the act of measurement gives reality to such a property as direction of motion. In fact, attempts have been made to show that the concept of quantum measurement involves logical paradoxes. One such attempt was made by Erwin Schrödinger, one of the founders of quantum mechanics.

In 1935, Schrödinger published a paper in the German scientific journal *Die Naturwissenschaften* in which he attempted to show that the Copenhagen interpretation was paradoxical. Schrödinger asked his readers to imagine that a cat had been placed in a closed steel chamber containing a small amount of radioactive material. Just enough of this substance was presumed to be present that the chance that one radioactive nucleus would decay in an hour was precisely one-half. Schrödinger imagined, also, that the box contained a Geiger counter that would detect the decay, if one occurred. Finally, the Geiger counter was imagined to be connected to an electrical circuit that would electrocute the cat* if the decay was registered.

According to Schrödinger, acceptance of the Copenhagen interpretation implied that until an observer opened the box to see whether or not the cat had been electrocuted, the animal could not be considered to be either alive or dead. Until an observation was made, the quantum states "cat alive" and "cat dead" existed with equal probability. It was only when the experimenter looked that one of them assumed objective reality. In Schrödinger's opinion, such a conclusion was absurd.

One might object that the cat would surely know whether or not it had been electrocuted, and that consequently one need not speak of the cat as existing in two simultaneous probability states if one wants to adhere to the Copenhagen outlook. Indeed, most physicists would interpret Schrödinger's thought experiment in this way. They would say that one or the other of the quantum probabilities became reality as soon as it was registered in the macroscopic world. In this case, the registration would take place when the emitted particle entered (or did not enter) the Geiger counter.

But this is not the only possible interpretation of the fate of Schrödinger's cat. According to John von Neumann, the Hungarian mathematician who published an important study of the mathematical foundations of quantum mechanics in 1932, it was impossible to formulate a consistent interpretation of the theory without reference to human consciousness. In recent years, this idea has been elaborated upon by the Hungarian-American physicist Eugene Wigner, who has applied it to Schrödinger's imaginary experiment.

According to Wigner, no measurement can be said to have been completed until the result enters into the mind of a conscious being. He claims that until this happens, nothing exists except an ensemble of quantum probabilities. Thus, in Wigner's view, the cat is half-alive and half-dead until a human being opens

---

* Nowadays, most of the published descriptions of this thought experiment replace the electric circuit with a canister of poison gas. Apparently, it is thought more humane to kill Schrödinger's cat with gas than with electricity.

the box to see whether or not it has been electrocuted. Even filming the cat would not cause it to assume one state or another. In such a case, the cat would again be half-alive and half-dead until a human being decided to view the film.

It appears that the acceptance of Wigner's viewpoint would imply that the universe could not really exist if there was no one here to observe it. Or at least it could exist only in some vague, probabilistic way. Hence it is not surprising that Wigner's arguments have been widely criticized. Even advocates of the Copenhagen interpretation, which requires that our ideas about the existence of reality be modified, are generally not willing to discard objectivity entirely and to make so much dependent upon the consciousness of the observer.

A somewhat different point of view is taken by John Archibald Wheeler. Wheeler stresses that "no elementary [quantum] phenomenon is a phenomenon until it is a registered phenomenon." However, he points out that "registered" can be taken to mean "brought to a close by an irreversible act of amplification" (an example of such an irreversible act would be the detection of a decay particle by a Geiger counter). According to Wheeler, the use of the term "registered" is preferable to "observed" because it implies no reference to consciousness.

Yet Wheeler reaches conclusions that are not so very different from those of Wigner. In fact, Wheeler claims that the universe cannot be said to exist independently of acts of registration. "The past," he says, "has no existence until it is recorded in the present." He comes to the conclusion that the universe could not have come into existence 15 billion years ago if it did not have the potential for containing conscious observers. According to Wheeler, acts of "observer-participancy" give "tangible 'reality' to the universe not only now but back to the beginning." Furthermore, he says, one can hope that we will someday be able to derive the structure of quantum mechanics from the requirement that everything have a way to come into being.

An interesting anticipation of some of Wheeler's ideas can be found in the writing of the German physiologist Paul Jensen. In 1934, a few years after the Copenhagen interpretation was formulated, Jensen argued that if one assumed that the properties of quantum objects were dependent upon observation, then any physical state of affairs would be dependent upon observation also. According to Jensen, if one took the reasoning upon which the Copenhagen interpretation was based and followed it to its logical conclusion, then one would have to infer that no objective set of facts could exist independently of the observations of conscious observers.

Jensen regarded this as a criticism of the Copenhagen interpretation. Wheeler, on the other hand, would reach a somewhat different conclusion. He would agree that acceptance of the Copenhagen interpretation implies such an outlook. However, he would say that we must therefore accept the fact that reality is, in some sense, dependent upon the existence of conseiousness.

It is possible to argue that the Copenhagen interpretation does not require acceptance of such views as those of Wigner and Wheeler. However, one cannot deny the fact that their conclusions are at least consistent with the Copenhagen outlook. Any theory that denies objective reality to some of the properties of quantum objects must raise questions about the role played by human consciousness.

After all, when we say that a quantity has "objective reality," we mean that its properties are independent of our perceptions. Objectively real objects are ones that will behave in the same manner whether we observe them or not. If the outlooks developed by physicists like Wigner and Wheeler seem a bit metaphysical, that is only to be expected. After all, the Copenhagen interpretation is basically a philosophical outlook. Its proponents take the point of view that our understanding of what we mean by "reality" must be modified when the term is applied to events in the quantum world.

If one is uncomfortable with these kinds of "reality," there are a number of alternatives that have already been discussed. Since there is no evidence that forces us to favor one interpretation of quantum mechanics over another, one is free to believe that there are an enormous number of parallel universes (the many-worlds interpretation) if one wishes. Or one can believe in the existence of nonlocal hidden variables, and maintain that quantum events which lie at great distances from one another may influence each other through a kind of quantum action-at-a-distance.

The question of which interpretation of quantum mechanics is most likely to be correct is not one that present-day science can answer. If we want to know what physical reality is "really" like, we must fall back on our own philosophical prejudices and pre-conceptions.

It seems natural to ask what, if anything, all this implies about the reality of the phenomena that have been discussed in this book, such as quarks, virtual particles, unified forces, ten- or eleven-dimensional space-time, and the like. If there are reasons to doubt the objective reality of certain properties of a much-studied particle like the electron, to what extent should we attribute reality to the more abstract concepts of contemporary physics?

The only answer that one can give is, It depends upon your philosophical outlook. One may choose to believe, as Stephen Hawking does, that we are on the verge of attaining complete knowledge of the fundamental laws of nature. One may believe, as most physicists probably do, that the best we can do is to build models, and that quarks, virtual particles, and so on, are constructs that we create. Finally, one can adopt an even more relativistic point of view, one that has been espoused by John Archibald Wheeler.

According to Wheeler, we live in a universe that is ruled by chaos and chance. He says that the laws of physics themselves may come about as the result of interactions between the universe and its participant-observers. We do not simply observe the laws of nature, he claims; there is also a sense in which we create them. Wheeler sees physics as a "meaning circuit." In his view, meaning is created when we put questions to nature, and when nature responds by producing observable phenomena in reply.

Many of Wheeler's pronouncements have a tendency to sound vague and mystical. However, as Freeman Dyson points out, this does not necessarily imply that we should not take them seriously. "We should have learned by now," Dyson says, "that ideas that appear at first sight to be vague and mystical sometimes turn out to be true."

Questions about the nature of reality can be subtle indeed. It can be extraordinarily difficult to state just what kinds of "reality" should be attributed to the concepts of modern physics. Eugene Wigner has told a little parable which illustrates this point.

Wigner begins by noting that, at one time, the question of

whether a magnetic field could exist in a vacuum was one that was hotly debated by physicists. Today, Wigner goes on, most physicists would say that such a thing does exist. The concept of a magnetic field enters their calculations, and they can explain their calculations and conclusions to others more fully if they assume that the field is real. According to Wigner, "the reality of the magnetic field in vacuum consists of the usefulness of the magnetic field concept . . . both for our own thinking and for communicating with others."

However, Wigner goes on, such a kind of reality is not absolute, but only relative. In a gathering of physicists, he says, the magnetic field is taken to be real. However, he goes on, "If I were cast upon an island abounding with poisonous snakes and had to defend my life against them, the reality of the magnetic field would fade, at least temporarily, into insignificance."

Wigner's purpose in telling this parable is to emphasize his belief that everything except our sensations and the contents of our consciousness is a construct. His purpose in doing this is to lend support to his idea that consciousness is an important element in the interpretation of quantum mechanics. It is not necessary to follow the road that Wigner travels if one does not wish to. However, it is difficult to read his anecdote without feeling that it has something important to say about the nature of physical reality.

I must admit that I find Wigner's pragmatic view to be quite congenial. There is something very appealing about the idea that the concepts of physics that are "real" are the ones that are useful. I don't know whether or not quarks "really exist." But if the concept of the quark gives us a way of understanding experimental data, and of predicting the existence of new phenomena, then perhaps quarks have reality enough. I, for one, fail to see what more we could demand.

Bibliography

Amaldi, Ginestra. *The Nature of Matter*. Chicago: University of Chicago Press, 1982.

Asimov, Isaac. *The Neutrino*. Garden City, N.Y.: Doubleday, 1966.

Barrow, John D., and Joseph Silk. *The Left Hand of Creation*. New York: Basic Books, 1983.

Bernstein, Jeremy. *Einstein*. Harmondsworth, England: Penguin, 1976.

Blumenthal, George R., et al. "Formation of Galaxies and Large-Scale Structures with Cold Dark Matter." *Nature* 311 (1984), pp. 517–25.

Bohr, Niels. *Atomic Theory and the Description of Nature*. New York: Macmillan, 1934.

Born, Max. *Physics in My Generation*. New York: Springer-Verlag, 1969.

———. *The Restless Universe*. New York: Dover, 1951.

Brockman, John. *Afterwords*. Garden City, N.Y.: Doubleday-Anchor, 1973.

Canuto, V., and S. H. Hsieh. "Case for an Open Universe." *Physical Review Letters* 44 (1980), pp. 695–98.

Cartwright, Nancy. *How the Laws of Physics Lie*. Oxford: Clarendon, 1983.

Clark, Ronald W. *Einstein*. New York: Avon, 1972.

Cline, Barbara Lovett. *The Questioners*. New York: Thomas Y. Crowell, 1965.

Close, Frank E. "Weighing Neutrinos: A Small Wait for a Small Weight?" *Nature* 289 (1981), pp. 747–48.

Crenmer, E., B. Julie, and J. Schenk. "Supergravity Theory in 11 Dimensions." *Physics Letters* 76B (1978), pp. 409–12.

Cropper, William H. *The Quantum Physicists*. New York: Oxford University Press, 1970.

d'Abro, A. *The Rise of the New Physics*. 2 vols. New York: Dover, 1952.

Davies, Paul. *The Accidental Universe*. Cambridge: Cambridge University Press, 1982.

———. *The Edge of Infinity*. New York: Simon & Schuster, 1981.

———. *The Forces of Nature*. Cambridge: Cambridge University Press, 1979.

———. "How Special Is the Universe?" *Nature* 249 (1974), pp. 208–9.

———. "The Inflationary Universe." *The Sciences* 23, No. 2 (March–April 1983), pp. 32–37.

———. *Superforce*. New York: Simon & Schuster, 1984.

De Rújula, A., and S. L. Glashow. "Galactic Neutrinos and uv Astronomy." *Physical Review Letters* 45 (1980). pp. 942–44.

———. "Neutrino Weight Watching." *Nature* 286 (1980), pp. 755–56.

d'Espagnat, Bernard. *Conceptual Foundations of Quantum Mechanics*. 2nd ed. Reading, Mass.: W. A. Benjamin, 1976.

———. "The Quantum Theory and Reality." *Scientific American* 241, No. 5 (November 1979), pp. 158–81.

de Vaucouleurs, G. "The Cosmological Distance Scale: A Comparison of the Approaches to the Hubble Constant." *Annals of the New York Academy of Sciences* 375 (1981), pp. 90–122.

DeWitt, Bryce S. "Quantum Gravity." *Scientific American* 249, No. 6 (December 1983), pp. 112–29.

DeWitt, Bryce S., and Neill Graham, eds. *The Many-Worlds Interpretation of Quantum Mechanics*. Princeton: Princeton University Press, 1973.

Dirac, P. A. M. *Directions in Physics*. New York: Wiley, 1978.

———. *The Principles of Quantum Mechanics*. 4th ed. (revised). Oxford: Clarendon, 1967.

Dodd, J. E. *The Ideas of Particle Physics*. Cambridge: Cambridge University Press, 1984.

Dodd, James. "Universal Supersymmetry." *New Scientist* 83 (1979).

Drake, Stillman. *Galileo*. New York: Hill & Wang, 1980.

Drell, Sidney D. "When Is a Particle?" *American Journal of Physics* 46 (1978), pp. 597–606.

Eddington, Arthur. *The Nature of the Physical World*. Ann Arbor: University of Michigan Press, 1958.

———. *The Philosophy of Physical Science*. Ann Arbor: University of Michigan Press, 1958.

Edmunds, M. G. "Open Debate." *Nature* 288 (1980), pp. 431–32.

Einstein, Albert. *Investigations on the Theory of the Brownian Movement*. New York: Dover, 1956.

———. *The Meaning of Relativity*, 5th ed. Princeton: Princeton University Press, 1956.

Einstein, Albert, and Leopold Infeld. *The Evolution of Physics*. New York: Simon & Schuster, 1938.

Einstein, Albert, et al. *The Principle of Relativity*. New York: Dover, 1952.

Ellis, G. F. R. "Limits to Verification in Cosmology." *Annals of the New York Academy of Sciences* 336 (1980), pp. 130–60.

Ellis, John. "Hopes Grow for Supersymmetry." *Nature* 313 (1985), pp. 626–27.

Faber, S. M., and J. S. Gallagher. "Masses and Mass-to-Light Ratios of Galaxies." *Annual Review of Astronomy and Astrophysics* 17 (1979), pp. 135–87.

Feinberg, Gerald. *What Is the World Made Of?* Garden City, N.Y.: Doubleday-Anchor, 1978.

———. *Solid Clues*. New York: Simon & Schuster, 1985.

Ferris, Timothy. *The Red Limit*. New York: Bantam, 1979.

Feynman, Richard. *The Character of Physical Law*. Cambridge, Mass.: MIT Press, 1965.

Freedman, Daniel Z., and Peter Niewenhuizen. "The Hidden Dimensions of Spacetime." *Scientific American* 252, No. 3 (March 1985), pp. 74–81.

Freund, Peter G. O., and Mark A. Rubin. "Dynamics of Dimensional Reduction." *Physics Letters* 97B (1980), pp. 233–35.

Fritzsch, Harald. *The Creation of Matter*. New York: Basic Books, 1984.

Gale, George. "Science and the Philosophers." *Nature* 312 (1984), pp. 491–95.

Gamow, George. *Thirty Years That Shook Physics*. New York: Doubleday-Anchor, 1966.

Gardner, Martin. "Can Time Go Backward?" *Scientific American* 309, No. 1 (January 1967), pp. 98–108.

Geroch, Robert. *General Relativity from A to B*. Chicago: University of Chicago Press, 1978.

Gibbons, G. W., S. W. Hawking, and S. T. C. Siklos. *The Very Early Universe*. Cambridge: Cambridge University Press, 1983.

Gott, J. Richard III, et al. "An Unbound Universe?" *Astrophysical Journal* 194 (1974), pp. 543–53.

Glashow, Sheldon Lee. "Grand Unification: Tomorrow's Physics." *New Scientist* 87 (1980), pp. 869–72.

Goldberg, Stanley. *Understanding Relativity*. Boston: Birkhäuser Verlag, 1984.

Graves, John Cowperthwaite. *The Conceptual Foundations of Contemporary Relativity Theory*. Cambridge, Mass.: MIT Press, 1977.

Green, Michael B. "Unification of Forces and Particles in Superstring Theories." *Nature* 314 (1985), pp. 409–14.

Gribbin, John. *In Search of Schrödinger's Cat*. New York: Bantam, 1984.

Guillemin, Victor. *The Story of Quantum Mechanics*. New York: Scribner, 1968.

Guth, Alan H., Kerson Huang, and Robert L. Jaffe, eds. *Asymptotic Realms of Physics*. Cambridge, Mass.: MIT Press, 1983.

Guth, Alan H., and Paul J. Steinhardt. "The Inflationary Universe." *Scientific American* 250, No. 5 (May 1984), pp. 116–28.

Hall, A. Rupert. *From Galileo to Newton*. New York: Dover, 1981.

Halzen, Francis, and Alan D. Martin. *Quarks and Leptons*. New York: Wiley, 1984.

Hanson, Norwood Russell. *The Concept of the Positron*. Cambridge: Cambridge University Press, 1963.

———. *Patterns of Discovery*. Cambridge: Cambridge University Press, 1972.

Harman, P. M. *Energy, Force and Matter*. Cambridge: Cambridge University Press, 1982.

Harré, Rom. *Great Scientific Experiments*. Oxford: Clarendon, 1983.

Hawking, Stephen. *Is the End in Sight for Theoretical Physics?* Cambridge: Cambridge University Press, 1980.

Hawking, Stephen, and W. Israel, eds. *General Relativity*. Cambridge: Cambridge University Press, 1979.

Heisenberg, Werner. *The Physicist's Conception of Nature*. Westport, Conn.: Greenwood Press, 1970.

———. *Physics and Beyond*. New York: Harper & Row, 1971.

———. *Physics and Philosophy*. New York: Harper & Row, 1962.

Hermann, Armin. *The Genesis of Quantum Theory (1899–1913)*. Cambridge, Mass.: MIT Press, 1971.

Hoffman, Banesh. *Albert Einstein*. New York: New American Library, 1973.

Hubble, Edwin. *The Realm of the Nebulae*. New Haven: Yale University Press, 1936.

Hund, Friedrich. *The History of Quantum Theory*. New York: Barnes & Noble, 1974.

Hut, Piet, and Simon D. M. White. "Can a Neutrino-Dominated Universe Be Rejected?" *Nature* 310 (1984), pp. 637–40.

"In Search of Physical Reality." *Nature* 313 (1985), p. 438.

Jammer, Max. *Concepts of Force*. Cambridge, Mass.: Harvard University Press, 1957.

———. *Concepts of Mass.* Cambridge, Mass.: Harvard University Press, 1961.

———. *The Conceptual Development of Quantum Mechanics.* New York: McGraw-Hill, 1966.

———. *The Philosophy of Quantum Mechanics.* New York: Wiley, 1974.

Jeans, James. *Physics and Philosophy.* New York: Dover, 1981.

Kazanas, Demosthenes, David N. Schramm, and Kem Hainebach. "A Consistent Age for the Universe." *Nature* 274 (1978), pp. 672–73.

Kline, Morris. *Mathematics: The Loss of Certainty.* New York: Oxford University Press, 1980.

Kolb, Edward W., David Seckel, and Michael S. Turner, "The Shadow World of Superstring Theories." *Nature* 314 (1985), pp. 415–19.

Kuhn, Thomas S. *The Copernican Revolution.* Cambridge, Mass.: Harvard University Press, 1957.

———. *The Structure of Scientific Revolutions.* Chicago: University of Chicago Press, 1970.

Lindsay, Robert Bruce. *The Nature of Physics.* Providence: Brown University Press, 1968.

Llewellyn-Smith, C. H. "Opportunities in Particle Physics." *Nature* 312 (1984), pp. 588–92.

MacDonald, D. K. C. *Faraday, Maxwell and Kelvin.* Garden City, N.Y.: Doubleday-Anchor, 1984.

Maddox, John. "Dispute over Scale of Universe." *Nature* 307 (1984), p. 313.

———. "How Special Is Special Relativity?" *Nature* 313 (1985), p. 429.

———. "Hunting for the Missing Mass." *Nature* 310 (1984), p. 627.

———. "The Importance of Impatience." *Nature* 314 (1985), p. 309.

Margenau, Henry. *The Nature of Physical Reality.* New York: McGraw-Hill, 1950.

Mehra, Jagdish, ed. *The Physicist's Conception of Nature.* Dordrecht, Holland: D. Reidel, 1973.

Miller, Arthur I. *Imagery in Scientific Thought.* Boston: Birkhäuser Verlag, 1984.

Misner, Charles W., Kip S. Thorne, and John Archibald Wheeler. *Gravitation.* San Francisco: Freeman, 1973.

Moore, Ruth. *Niels Bohr.* New York: Knopf, 1966.

Morris, Richard. *Dismantling the Universe.* New York: Simon & Schuster, 1983.

———. *Time's Arrows.* New York: Simon & Schuster, 1985.

Motz, Lloyd. *The Universe.* New York; Scribner, 1975.

Mulvey, J. H., ed. *The Nature of Matter.* Oxford: Clarendon, 1981.

Nelson, Philip. "Naturalness in Theoretical Physics." *American Scientist* 73, No. 1 (January–February 1985), pp. 60–67.

Newton, Isaac. *Mathematical Principles*. Berkeley: University of California Press, 1946.

Pagels, Heinz R. *The Cosmic Code*. New York: Simon & Schuster, 1982.

――――. *Perfect Symmetry*. New York: Simon & Schuster, 1985.

Pais, Abraham. *"Subtle Is the Lord. . . ."* Oxford: Clarendon, 1982.

Particles and Fields: Readings from Scientific American. San Francisco: Freeman, 1980.

Petersen, Aage. *Quantum Physics and the Philosophical Tradition*. Cambridge, Mass.: MIT Press, 1968.

Pickering, Andrew. *Constructing Quarks*. Chicago: University of Chicago Press, 1984.

Polanyi, Michael. *Personal Knowledge*. New York: Harper Torchbooks, 1964.

Popper, Karl R. *The Logic of Scientific Discovery*. New York: Harper Torchbooks, 1968.

Przibram, Karl, ed. *Letters on Wave Mechanics*. New York: Philosophical Library, 1967.

Quigg, Chris. "Elementary Particles and Forces." *Scientific American* 252, No. 4 (April 1985), pp. 84–95.

Raychaudhur, A. K. *Theoretical Cosmology*. Oxford: Clarendon, 1979.

Reichenbach, Hans. *The Philosophy of Space and Time*. New York; Dover, 1957.

――――. *The Rise of Scientific Philosophy*. Berkeley: University of California Press, 1951.

Robinson, Arthur L. "Quantum Mechanics Passes Another Test." *Science* 217 (1982), pp. 435–36.

Russell, Bertrand. *Human Knowledge*. New York: Simon & Schuster, 1948.

――――. *The Scientific Outlook*. New York: Norton, 1962.

Sachs, Mendel. *The Field Concept in Contemporary Science*. Springfield, Ill.: Charles C Thomas, 1973.

Sandage, Allan R. "Cosmology: A Search for Two Numbers." *Physics Today* 23, No. 2 (1970), pp. 34–41.

Scheibe, Erhard. *The Logical Analysis of Quantum Mechanics*. Oxford: Pergamon Press, 1973.

Schlegel, Richard. *Superposition and Interaction*. Chicago: University of Chicago Press, 1980.

Schonland, Sir Basil. *The Atomists*. Oxford: Clarendon, 1968.

Schramm, David N., and Robert V. Wagoner. "Element Production in the Early Universe." *Annual Reviews of Nuclear Science* 27 (1977), pp. 37–74.

Segrè, Emilio. *From Falling Bodies to Radio Waves*. New York: Freeman, 1984.

――――. *From X-Rays to Quarks*. San Francisco: Freeman, 1980.

Shipman, Harry L. *Black Holes, Quasars and the Universe.* 2nd ed. Boston: Houghton Mifflin, 1980.

Silk, Joseph. *The Big Bang.* San Francisco: Freeman, 1980.

———. "Cosmologists in the Dark." *Nature* 311 (1984), pp. 508–9.

Stapp, Henry Pierce. "The Copenhagen Interpretation." *American Journal of Physics* 60 (1972), pp. 1098–1116.

Sunyaev, R. A., and Ya. B. Zel'dovich. "Microwave Background Radiation as a Probe of the Contemporary Structure and History of the Universe." *Annual Review of Astronomy and Astrophysics* 18 (1980), pp. 537–60.

Tayler, R. J. "Brown Dwarfs and Hidden Mass." *Nature* 316 (1985), pp. 19–20.

ter Haar, D. *The Old Quantum Theory.* Oxford: Pergamon Press, 1967.

Thomsen, Dietrick E. "A Closed Universe May Be Axionomatic." *Science News* 125 (1984), pp. 396–97.

———. "The New Inflationary Nothing Universe." *Science News* 123 (1983), pp. 108–9.

Thomson, Sir George Paget. *J. J. Thomson and the Cavendish Laboratory.* London: Nelson, 1964.

Toulmin, Stephen, and June Goodfield. *The Architecture of Matter.* Chicago: University of Chicago Press, 1982.

———. *The Fabric of the Heavens.* New York: Harper, 1961.

Trefil, James E. *From Atoms to Quarks.* New York: Scribner, 1980.

———. *The Moment of Creation.* New York: Scribner, 1983.

Trigg, George L. *Crucial Experiments in Modern Physics.* New York: Van Nostrand, 1971.

Turner, Michael S., and David N. Schramm. "Cosmology and Elementary-Particle Physics." *Physics Today* 32, No. 9 (September 1979), pp. 42–48.

Wald, Robert M. *Space, Time and Gravity.* Chicago: University of Chicago Press, 1977.

Waldrop, M. Mitchell. "Before the Beginning." *Science 84* 5, No. 1 (January–February 1984), pp. 44–51.

———. "The New Inflationary Universe." *Science* 219 (1983), pp. 375–77.

———. "New Light on Dark Matter?" *Science* 224 (1984), pp. 971–73.

Walgate, Robert. "Squarks and Strings More Real." *Nature* 313 (1985), p. 9.

Watzlawick, Paul. *How Real Is Real?* New York: Vintage, 1977.

Wayson, W. H. *Understanding Physics Today.* Cambridge: Cambridge University Press, 1963.

Weinberg, Steven. *The First Three Minutes.* New York: Basic Books, 1977.

———. "The Search for Unity: Notes for a History of Quantum Field Theory." *Daedalus* 106, No. 4 (Fall 1973), pp. 17–35.

Weisskopf, Victor F. *Physics in the Twentieth Century*. Cambridge, Mass.: MIT Press, 1972.

Wheeler, John Archibald. "Bits, Quanta, Meaning." In A. Giovanni, F. Mancini, and M. Marinaro, eds. *Problems in Theoretical Physics*. Salerno: University of Salerno Press, 1984.

Wheeler, John Archibald, and Wojciech Hubert Zurek, eds. *Quantum Theory and Measurement*. Princeton: Princeton University Press, 1983.

Wigner, Eugene P. *Symmetries and Reflections*. Cambridge, Mass.: MIT Press, 1970.

Witter, Edward. "Search for a Realistic Kaluza-Klein Theory." *Nuclear Physics* B186 (1981), pp. 412–28.

Woolf, Harry, ed. *Some Strangeness in the Proportion*. Reading, Mass.: Addison-Wesley, 1980.

Index

241